Ionic and Electrochemical Equilibria

Chemical Thermodynamics Set

coordinated by
Michel Soustelle

Volume 6

Ionic and Electrochemical Equilibria

Michel Soustelle

iSTE

WILEY

First published 2015 in Great Britain and the United States by ISTE Ltd and John Wiley & Sons, Inc.

ISTE Ltd
27-37 St George's Road
London SW19 4EU
UK

www.iste.co.uk

John Wiley & Sons, Inc.
111 River Street
Hoboken, NJ 07030
USA

www.wiley.com

Library of Congress Control Number: 2016936176

British Library Cataloguing-in-Publication Data
A CIP record for this book is available from the British Library
ISBN 978-1-84821-869-7

Contents

Preface

This book – an in-depth examination of chemical thermodynamics – is written for an audience of engineering undergraduates and Masters students in the disciplines of chemistry, physical chemistry, process engineering, materials, etc., and doctoral candidates in those disciplines. It will also be useful for researchers at fundamental- or applied-research labs, dealing with issues in thermodynamics during the course of their work.

These audiences will, during their undergraduate degree, have received a grounding in general thermodynamics and chemical thermodynamics, which all science students are normally taught. This education will undoubtedly have provided them with the fundamental aspects of macroscopic study, but usually the phases discussed will have been fluids exhibiting perfect behavior. Surface effects, the presence of an electrical field, real phases, the microscopic aspect of modeling, and various other aspects, are hardly touched upon (if at all) during this early stage of an academic career in chemical thermodynamics.

This series, which comprises 7 volumes, and which is positioned somewhere between an introduction to the subject and a research thesis, offers a detailed examination of chemical thermodynamics that is necessary in the various disciplines relating to chemical- or material sciences. It lays the groundwork necessary for students to go and read specialized publications in their different areas. It constitutes a series of reference books that touch on all of the concepts and methods. It discusses both scales of modeling: microscopic (by statistical thermodynamics) and macroscopic, and illustrates the link between them at every step. These models are then used in the study of solid, liquid and gaseous phases, either of pure substances or comprising several components.

The different instalments in this series deal with the following subjects:

– single-phase macroscopic and microscopic modeling tools: application to gases;

– modeling of liquid phases;

– modeling of solid phases;

– chemical equilibrium states;

– phase transformations;

– electrolytes and electrochemical thermodynamics;

– thermodynamics of surfaces, capillary systems and phases of small dimensions.

Appendices in each volume give an introduction to the general methods used in the text, and offer reminders and additional tools.

This series owes a great deal to the feedback, comments and questions from all my students at the *Ecole national esupérieure des mines* (engineering school) in Saint Etienne who have “endured” my lecturing in thermodynamics for many years. I am very grateful to them, and also thank them for their stimulating attitude. This work is also the fruit of numerous discussions with colleagues who teach thermodynamics in the largest establishments – particularly in the context of the group “Thermodic”, founded by Marc Onillion. My thanks go to all of them for their contributions and kindness.

This sixth volume is made up of two parts: one devoted to ionic equilibria and the other to electrochemical thermodynamics.

In the first part, we discuss the concepts of dissociation of electrolytes and the phenomena of solvation in the different types of solvents – aqueous and non-aqueous. Next, the different families of ionic equilibria are studied, in turn looking at acid–base equilibria, the equilibria of complex formation, redox reactions and equilibria of precipitation. In each case, we examine the phenomena in both an aqueous and a non-aqueous medium. Solid electrolytes are also touched upon.

Part 2 is dedicated to electrochemical thermodynamics with the involvement of charges in electrical fields. A general approach is used to define the electrochemical values, such as the electrochemical potential of a species, the electrochemical Gibbs energy of a system, etc. Then, two different types of electrochemical systems are studied – first, electrodes with the corresponding reactions for the different types, and then galvanic. Applications of the measurements to galvanic cells are described, with a view to determining various thermodynamic values.

Finally, this second part closes with the study of potential/pH diagrams and their generalization in potential/pX diagrams, in aqueous- or non-aqueous media.

Michel SOUSTELLE
Saint-Vallier
March 2016

Notations and Symbols

A:	area of a surface or an interface.
$A_{\mathrm{H}}^{(12)}$:	Hamaker constant between two media, 1 and 2.
$\mathscr{A}$:	affinity
$\tilde{\mathscr{A}}$:	electrochemical affinity.
A_M:	molar area.
A_m:	molecular area.
a:	pressure of cohesion of a gas or radius of the elementary cell of a liquid.
A, B, …:	components of a mixture.
C:	concentration or plot concentration of a potential/pH diagram.
C_P^{xs}:	excess molar specific heat capacity at constant pressure.
C_i:	molar concentration (or molarity) of component i.
$C_{\pm}$:	mean concentration of ions in an ionic solution.
C_V, C_P:	specific heat capacity at constant volume and pressure.
c:	capacity of a condenser or number of independent components.
D:	dielectric constant of the medium.

d:	distance between two liquid molecules.
d_eS:	exchange of entropy with the outside environment.
d_i:	degree of oxidation i of an element A.
d_iS:	internal entropy production.
E:	energy in the system.
E^0:	standard electrical potential or standard electromotive force of a cell.
E_{abs}:	reversible electrical voltage of an electrochemical cell.
$\mathcal{E}_p$:	set of variables with p intensive variables chosen to define a system.
e:	relative voltage of an electrode.
e^0:	standard electrical potential (or normal voltage) of an electrode.
e_0:	equi-activity- or equiconcentration voltage of an electrode.
e_{abs}:	absolute voltage of an electrode.
F:	free energy.
$\tilde{F}$:	electrochemical free energy.
F_m:	molar free energy.
$\mathfrak{F}$:	faraday (unit).
$\tilde{G}_\sigma$:	electro-capillary Gibbs energy.
$\tilde{G}$:	electrochemical Gibbs energy.
G_m:	molar Gibbs energy.
g:	osmotic coefficient.
g_i^0:	molar Gibbs energy of the pure component i.
H_0:	Hammett acidity function
H_T^0:	standard molar enthalpy of formation at temperature T.

$H, \overline{H}_i$: enthalpy, partial molar enthalpy of i.

$\tilde{H}$: electrochemical enthalpy.

h: stoichiometric coefficient of the protons in an electrochemical reaction.

h: Planck's constant.

h_i^0: molar enthalpy of the pure component i.

I: ionic strength of a solution of ions.

I_m: ionic strength expressed in terms of the molalities.

i: van 't Hoff factor.

K_{AX}: solubility product of the solid AX.

K_d: dissociation constant.

$K_r^{(c)}$: equilibrium constant relative to the concentrations.

$K_r^{(f)}$: equilibrium constant relative to the fugacities.

$K_r^{(P)}$: equilibrium constant relative to the partial pressures.

K_r: equilibrium constant.

K_s: solubility product.

$\mathrm{k_B}$: Boltzmann's constant.

M: molar mass.

m_s: mass of solutes in grams per kg of solvent.

m: total mass.

m_i: mass of component i.

N: number of components of a solution.

$\mathrm{N_a}$: Avogadro's number.

N_A: number of molecules of component A.

$n^{(\alpha)}$: total number of moles in a phase α.

P:	pressure of a gas.
P_i:	partial pressure of the component i.
p:	number of external physical variables.
Q_a:	reaction quotient in terms of activities.
Q_P:	heat of transformation at constant pressure; reaction quotient in terms of partial pressures.
Q_r:	reaction quotient of the transformation r.
R:	perfect gas constant.
r_A:	radius of the ionic atmosphere.
S:	oversaturation of a solution.
$\tilde{S}$:	electrochemical entropy.
s_i^0:	molar entropy of the pure component i.
T:	temperature
$\tilde{U}$:	internal electrochemical energy.
u_i^0:	molar internal energy of the pure component i.
$V, \overline{V}_i$:	volume, partial molar volume of i.
V_m:	molar volume.
v_i^0:	molar volume of the pure component i.
v:	quantum number of vibration.
w_i:	mass fraction of the component i.
$x_k^{(\alpha)}$:	molar fraction of the component k in the α phase.
x, y, z:	coordinates of a point in space.
x_i:	molar fraction of the component i in a solution.
$\langle y \rangle$:	mean value of y.
Y_i and X_i:	intensive and extensive conjugate variables.

y_i:	molar fraction of the component i in a gaseous phase.
α:	dissociation coefficient of a weak electrolyte or polarizability of a molecule.
α_a:	apparent dissociation coefficient of a weak electrolyte.
$\Gamma(\mathcal{E}_P)$:	characteristic function having the set $\mathcal{E}_P$ as canonical variables.
Γ:	characteristic function.
γ:	activity coefficient of the component i irrespective of the reference state.
γ_0:	activity coefficient of a solvent.
γ_i:	activity coefficient of the species i.
$\gamma_i^{(I)}$:	activity coefficient of component i in the pure-substance reference.
$\gamma_i^{(II)}$:	activity coefficient of component i in the infinitely-dilute-solution reference.
$\gamma_i^{(III)}$:	activity coefficient of component i in the molar-solution reference.
$\gamma_\pm$:	mean activity coefficient of the ions in an ionic solution.
γ_s:	activity coefficient of a solute.
$\Delta_r(A)$:	value of A associated with the transformation r.
ε:	electrical permittivity of the medium.
ε_0:	electrical permittivity of a vacuum.
λ_{0+}, λ_{0}:	equivalent ionic conductivities of the cation and the anion.
λ_A:	absolute activity of component A.
Λ:	equivalent conductivity of an electrolyte.
Λ_0:	limiting equivalent conductivity of an electrolyte.

μ_i:	chemical potential of component *i*, electrical dipolar moment of the molecule *i*.
$\mu_i^{(L)}, \mu_i^{(G)}$:	chemical potential of the component *i* in liquid and gaseous form, respectively.
$\tilde{\mu}$:	electrochemical potential.
$\nu_{k(\rho)}$:	algebraic stoichiometric number of component A_k in the reaction ρ.
ν_e:	stoichiometric coefficient of electrons in an electrochemical reaction.
ξ:	reaction progress.
Φ:	electrical potential.
Φ_i :	fugacity coefficient of component *i* in a gaseous mixture.
ϕ:	conductivity coefficient of a strong electrolyte or number of phases.
χ:	electrical conductivity.
Ψ_i :	electrostatic potential of the ionic atmosphere.
$\Psi(r)$:	electrostatic potential.

PART 1

Ionic Equilibria

1

Dissociation of Electrolytes in Solution

The dissociation of electrolytes – be it partial or total – in water releases ions, which lend the medium particular properties.

The ionic solution is characterized by the presence in the medium (generally a liquid) of ions carrying positive and negative charges, with the whole being electrically neutral. These ions may or may not be accompanied by:

– neutral dissolved molecules;

– molecules of solvent.

1.1. Strong electrolytes – weak electrolytes

Starting with neutral molecules in solid- or gaseous form, there are three main ways to obtain a liquid ionic solution: dissolution, solvolysis and melting.

1.1.1. *Dissolution*

When we place sodium chloride crystals in water, they dissolve according to the reaction:

$$NaCl_{(solid)} = Na^{+}_{(aqu)} + Cl^{-}_{(aqu)} \quad [1R.1]$$

In fact, the ionic solution obtained is the result of three phenomena: dissociation into ions, solvation of ions (in this case, hydration), which is the fixation of a certain number of polar molecules of solvent onto the ions and the separation of the charges of opposite signs because of the high electrical permittivity of the solvent.

1.1.2. *Solvolysis*

Solvolysis is the decomposition of a molecule by a solvent. In the case of water, we speak of hydrolysis. Take the example of gaseous hydrogen chloride composed of HCl molecules, whose reaction with water leads to the formation of ions by the following reaction, which is indeed a solvolysis:

$$HCl_{(gas)} + H_2O = H_3O^+_{(aqu)} + Cl^-_{(aqu)} \quad [1R.2]$$

The result is the presence of ions, which are also solvated and separated from one another for the same reasons as in dissolution.

1.1.3. *Melting*

Raising the temperature of a solid such as sodium chloride leads to its melting, which leads to the dissociation into ions, according to the reaction:

$$NaCl_{(solid)} = Na^+_{(liq)} + Cl^-_{(liq)} \quad [1R.3]$$

We again obtain a solution of ions (and neutral molecules), which are obviously not solvated, because the solution does not contain any solvent in the true sense of the word.

When a solution is obtained by one of the methods described above, we obtain a solution with multiple interactions between the ions, which can be described in one of two ways:

– a complex solution of ions with activity coefficients using a more or less elaborate model;

– a quasi-chemical model using the model of associated solutions, which leads us to divide the species in the solution into two categories:

- neutral associated molecules,

- ions.

Remember that the associated solution model consists of replacing a non-perfect solution of ions or molecules, generally complex, with a less complex solution (a perfect solution, a dilute ideal solution or a relatively-simple model), formed of the same ions and accompanied by ionic or molecular associated species at equilibrium with the ions.

Depending on the nature of the species in question, we are then led to distinguish two types of solutions:

– solutions which practically contain only ions;

– solutions which, alongside the ions, contain a not-insignificant amount of non-dissociated neutral molecules. These molecules may be molecules of the solvent or of a solute.

If the amount of non-dissociated neutral molecules is negligible in comparison to that of the dissociated molecules, we say that we have a strong electrolyte; such is the case of the aqueous solutions of sodium chloride and hydrogen chloride gas seen earlier. If, on the other hand, the number of molecules not dissociated is significant, we say that we are dealing with a weak electrolyte; such is the case of molten sodium chloride at a temperature a little above the melting point. It is also the case with the aqueous solution of ethanoic acid or ammonia, for example.

In practice, a strong electrolyte is an ionic solution whose formation reaction is complete toward the right; it no longer contains any neutral molecules. Meanwhile, a weak electrolyte is characterized by states of thermodynamic equilibrium between the ions and the neutral molecules – i.e. ultimately characterized by equilibrium constants.

In aqueous solution, practically all salts are strong electrolytes, whilst acids and bases are divided into strong acids and bases, on the one hand, and weak acids and bases, on the other.

1.2. Mean concentration and mean activity coefficient of ions

The methods for measuring the activity coefficients are unable to give us the activity coefficients of the individual ions, so it is useful to introduce, for an electrolyte $A_{\nu_+}B_{\nu_-}$, the idea of the *mean activity coefficient* which gives us the same Gibbs energy. We can show that this coefficient is defined by:

$$\gamma_\pm = \left(\gamma_+^{\nu_+}\gamma_-^{\nu_-}\right)^{1/(\nu_+ + \nu_-)} \qquad [1.1]$$

One might also define a *mean concentration* using a similar relation. If C is the molar concentration of the solute, the concentrations of the different ions (for entirely-dissociated strong electrolytes) will be:

$$C_+ = \nu_+ C \qquad [1.2a]$$

and

$$C_- = \nu_- C \qquad [1.2b]$$

and the mean concentration will be:

$$C_\pm = C\left(\nu_+^{\nu_+}\nu_-^{\nu_-}\right)^{1/(\nu_+ + \nu_-)} \qquad [1.3]$$

The mean activity coefficient obeys the same convention as the individual activity coefficients – generally convention (III) – but we know that in a dilute solution, the activity coefficients in conventions (II) and (III) are identical.

In particular, for a so-called 1–1 electrolyte, such as potassium chloride (the dissociation of the neutral molecule yields one anion and one cation), we have $\nu_+ = \nu_- = 1$ and the above expressions take the following forms:

$$C_+^{1/2} = C_-^{1/2} = C^{1/2} = C_\pm \qquad [1.4]$$

$$\gamma_\pm = \left(\gamma_+\gamma_-\right)^{1/2} \qquad [1.5]$$

1.3. Dissociation coefficient of a weak electrolyte

Consider an electrolyte A which dissociates according to reaction [1R.4], giving rise to ν_+ cations $A^{z_+ +}$ and ν_- anions $A^{z_- -}$:

$$A = \nu_+ A^{z+} + \nu_- A^{z- -} \qquad [1R.4]$$

Electrical neutrality must be preserved:

$$\nu_+ z_+ = \nu_- z_- \qquad [1.6]$$

The dissociation constant is defined as the equilibrium constant of [1R.4]; which is expressed on the basis of the activities of the species (ions and non-dissociated molecules):

$$K_d = \frac{\left|A^{z+}\right|^{\nu_+} \left|A^{z-}\right|^{\nu_-}}{|A|} \qquad [1.7]$$

We know that, in a dilute solution, the molar concentrations are practically equal to the ratio of the molar fractions to the volume molar of the solvent, generally the water. However, at ambient temperature, the molar volume of water is basically 1 kg/l. Thus, we keep the same equilibrium constant where the concentrations, expressed in moles/l, replace the molar fractions. In addition, if we separate the concentrations of the activity coefficients, we can write:

$$K_d = \frac{\gamma_{A^{z+}}^{\nu+} \gamma_{A^{z-}}^{\nu-}}{\gamma_A} \frac{\left[A^{z+}\right]^{\nu_+} \left[A^{z-}\right]^{\nu_-}}{[A]} = K_d^{(\gamma)} K_d^{(C)} \qquad [1.8]$$

Thus, relation [1.8] defines two pseudo-constants – one relative to the concentrations:

$$K_d^{(C)} = \frac{\left[A^{z_+}\right]^{\nu_+} \left[A^{z_-}\right]^{\nu_-}}{[A]} \qquad [1.9]$$

and the other relative to the activity coefficients:

$$K_d^{(\gamma)} = \frac{\gamma_{A^+}^{\nu_+} \gamma_{A^-}^{\nu_-}}{\gamma_A} \qquad [1.10]$$

If we bring in the mean activity coefficient of the ions, then by applying relation [1.1], we find:

$$K_d^{(\gamma)} = \frac{\gamma_{A^{z_+}}^{\nu_+} \gamma_{A^{z_-}}^{\nu_-}}{\gamma_A} = \frac{\gamma_\pm^{(\nu_+ + \nu_-)}}{\gamma_A} \qquad [1.11]$$

NOTE.– If we look at relation [1.7], it seems that the dissociation constants do not depend on the solvent. In reality, the reaction written in the form [1R.4] is not correct, because it ignores all the solvation processes which we shall discuss in Chapter 2, which yield the fact that the constant for equilibrium [1.7] truly depends on the solvent.

For weak electrolytes, we define the *dissociation coefficient* or *ionization coefficient* α by the fraction of the molecules of electrolyte that are actually dissociated into the solution.

If we begin with C_0 moles of the molecular compound A, the dissociation represented by the reaction [1R.4] gives us a residual concentration of A of $C_0(1\text{-}\alpha)$, a concentration of $A^{z_+ +}$ which is $C_0\nu_+\alpha$ and a concentration of $A^{z_- -}$ which is $C_0\nu_-\alpha$. The law of mass action in the form [1.8], replacing the concentrations of the different species with their values as a function of α, is written:

$$K_d = \frac{\gamma_\pm^{(\nu_+ + \nu_-)}}{\gamma_A} \frac{C_0^{(\nu_+ + \nu_- - 1)} \alpha^{(\nu_+ + \nu_-)} \nu_+^{\nu_+} \nu_-^{\nu_-}}{1\text{-}\alpha} \quad [1.12]$$

With very dilute solutions, the activity coefficients are equal to 1 and the law of mass action is expressed as a function of the dissociation coefficient thus:

$$K_d = K_d^{(C)} = \frac{C_0^{(\nu_+ + \nu_- - 1)} \alpha^{(\nu_+ + \nu_-)} \nu_+^{\nu_+} \nu_-^{\nu_-}}{1\text{-}\alpha} \quad [1.13]$$

Later on, we use expressions [1.12] and [1.13] to determine the dissociation constant of a weak electrolyte.

NOTE.– If the concentration C_0 tends toward zero, we can use relation [1.13], and we see that in order for the equilibrium constant to remain finite, the denominator must tend toward zero, and thus the dissociation coefficient α must tend toward 1. Hence, the dissociation of a weak electrolyte tends to be complete if dilution becomes infinite. In other words, at infinite dilution, weak electrolytes behave like strong electrolytes.

1.4. Conduction of electrical current by electrolytes

Electrolytic solutions containing electrically-charged ions conduct electricity which they are subjected to a potential difference – i.e. when the ions are placed in an electrical field. A portion I_+ of the intensity of the current is delivered by the cations, which move in the direction of the field; the other portion I_- is carried by the anions, which move in the opposite direction to the field. The total intensity of the current is the sum of the cationic and anionic contributions:

$$I = I_+ + I_- \qquad [1.14]$$

The study of the conductivity of electrolytes does not, strictly speaking, fall within the field of thermodynamics. Nonetheless, here, we shall discuss the essential elements that are necessary to make use of that conductivity to determine the dissociation coefficients.

1.4.1. *Transport numbers and electrical conductivity of an electrolyte*

We use the term *cationic transport number* to denote the portion of the current transported by the cations. It is defined by:

$$\tau_+ = \frac{I_+}{I_+ + I_-} \qquad [1.15]$$

In parallel, we define the *anionic transport number* as:

$$\tau_- = \frac{I_-}{I_+ + I_-} \qquad [1.16]$$

Of course, by virtue of relation [1.14], we have:

$$\tau_+ + \tau_- = 1 \qquad [1.17]$$

If we consider a cell containing the electrolytical solution, of length l and section area s, the resistance obeys the law:

$$R = \rho \frac{l}{s} = \frac{1}{\chi} \frac{l}{s} \qquad [1.18]$$

χ is the electrical conductivity of the electrolyte, and we deduce:

$$\chi = \frac{1}{R}\frac{l}{s} \tag{1.19}$$

We can write, in view of Ohm's law, that if Q is the quantity of electricity which has passed through the cell uniformly during the time t, the voltage U at the terminals of the cell is:

$$U = RI = \frac{RQ}{t} \tag{1.20}$$

Thus, by comparing with relation [1.19]:

$$\chi = \frac{Q}{ts\frac{U}{l}} \tag{1.21}$$

The conductivity thus appears as the quantity of electricity per second passing across a 1 cm^2 section with a potential drop of 1 v/cm. According to relation [1.19], it is expressed in $\Omega^{-1}cm^{-1}$.

NOTE.– Above, we chose commonly-used units. Obviously, in the international system of units (SI), conductivity is expressed in $\Omega^{-1}m^{-1}$.

1.4.2. *Equivalent conductivity and limiting equivalent conductivity of an electrolyte*

Experience tells us that the conductivity of a solution depends on the concentration of electrolyte which it contains. Thus, it has become common practice to express the conductivity in relation to the concentration – i.e. the amount of dissolved salt (in moles) per cm^3 of solution.

Thus, let C be the concentration of a solution in moles/l. Thus, the quantity per cm^3 would be $C/1000$, and we define the *equivalent conductivity* Λ as the ratio of the conductivity to the number of equivalents per cm^3:

$$\Lambda = \frac{1000\chi}{Cz} \tag{1.22}$$

The equivalent conductivity is expressed in Ω^{-1}moles^{-1}cm^{2}.

NOTE.– Sometimes, although it is not widely used, we encounter the definition of the *molar conductivity* as the ratio of the conductivity to the concentration expressed in moles per liter:

$$\Lambda_m = \frac{\chi}{C} \qquad [1.23]$$

This molar conductivity is expressed in $l.\Omega^{-1}$mole^{-1}cm^{-1}.

Experience shows us that the equivalent conductivity increases as the concentration decreases, tending toward a limit as the concentration tends toward zero (infinite dilution). We define the *limiting equivalent conductivity* Λ_0 as being the conductivity at infinite dilution. Thus, we write:

$$\Lambda_0 = \lim_{C \to 0} \Lambda \qquad [1.24]$$

1.4.3. *Ionic mobility*

We know that each ion, supposed to be punctual, with a charge *ze*, placed in an electrical field $\vec{E}$ experiences a force $\vec{F}$ such that:

$$\vec{F} = ze\vec{E} \qquad [1.25]$$

That force imbues the ion with a velocity $\vec{V}$ in the direction of $\vec{E}$ or the opposite direction, depending on whether it is a cation or an anion. In its motion, the ion encounters resistance, which slows it down. If the solution is sufficiently dilute for it not to be influenced at all by the other ions, it only experiences a counter force on the part of the solvent. This is known as the Stock force, and is proportional to its velocity, in accordance with:

$$\vec{F}' = 6\pi\eta\, r_A \vec{V} \qquad [1.26]$$

In this expression, η is the viscosity of the solvent. If I is the ionic strength of the medium and ε is the electrical permeability of the medium, r_A is the ionic radius defined by the relation:

$$r_A = \sqrt{\frac{1000\varepsilon k_B T}{2N_a e^2 I}} \qquad [1.27]$$

The action of the two forces in opposite directions lends the ion a limiting velocity $\vec{V}_0$ such that:

$$6\pi\eta\, r_A \vec{V}_0 = ze\vec{E} \qquad [1.28]$$

Consider the ratio $\vec{V}_0/\vec{E}$, written as u_{0+} or u_{0-}. This velocity per unit field strength, depending on whether it is a cation or an anion, is called the *ionic mobility of the cation or of the anion.* Thus, for the respective mobilities of the cation and the anion, we have:

$$u_{0+} = \frac{z_+ e}{6\pi\eta r_+} \qquad [1.29a]$$

and

$$u_{0-} = \frac{z_- e}{6\pi\eta r_-} \qquad [1.29b]$$

Mobility is expressed in $cm^2s^{-1}V^{-1}$. The mobilities of the different ions in water range between 2.10^{-4} and 10^{-3}, with the exception of those of the H^+ and OH^- ions, which are much higher, with 3.10^{-3} for the proton and 2.10^{-3} for the hydroxide ion.

The mobilities defined above were to be understood in a sufficiently-dilute (or infinitely-dilute) solution, so that the ion is influenced only by the solvent. In a less dilute solution, each ion is influenced by its neighbors, as it is surrounding by an ionic atmosphere whose electrical charge is of the opposite sign to its own. Whilst at rest, the two centers of symmetry – of the ion and of its ionic atmosphere – coincide; the same is no longer true when the ion is subjected to the electrical field. The ion is then subject to two additional forces of resistance:

– the relaxation of the ionic atmosphere due to the fact that the ion tends to move in one direction and its ionic atmosphere in the other direction;

– the electrophoretic effect: the counter-flow movement of the positive and negative ions increases the difficulty for the ions to move in the solution.

It follows that in a non-infinitely-dilute solution, the mobilities u_+ and u_- of the ions are less than their values observed at infinite dilution, i.e. at zero concentration, denoted by u_{0+} and u_{0-}.

Debye and Hückel, alongside Onsager, showed that the mobility at concentration C can be obtained by dividing the mobility at zero concentration by the same corrective term as that used for the activity coefficients in the Debye–Hückel model of a solution, meaning that, if we consider the Debye–Hückel limiting law (see section A.5, in the Appendix), we have:

$$u_+ = \frac{u_{0+}}{1 + Br_+\sqrt{I}} \qquad [1.30a]$$

and

$$u_- = \frac{u_{0-}}{1 + Br_-\sqrt{I}} \qquad [1.30b]$$

The coefficient B is always given by the expression:

$$B = 2.303\frac{e^2\sqrt{\dfrac{2\mathrm{N_a}e^2}{1000\,\varepsilon_0 D\mathrm{k_B}T}}}{2D\varepsilon_0\mathrm{k_B}T} \qquad [1.31]$$

where, in water, $B = 0.511\,\mathrm{l}^{1/2}\mathrm{mole}^{-0.5}$.

NOTE.– Relations [1.30a] and [1.30b] are valid within the same range of concentrations as the Debye–Hückel relation.

1.4.4. *Relation between equivalent conductivity and mobility – Kohlrausch's law*

Consider a solution of a completely ionized electrolyte (strong electrolyte) at a concentration C that is sufficiently low so that the mobilities of the ions are the limiting mobilities u_{0+} and u_{0-}. The cationic concentration is ν_+C, whilst that of the anions is ν_-C.

– the number of cations per cm^3 is, therefore: $\nu_+C/1000$;

– the number of anions per cm^3 is: $\nu_-C/1000$.

The numbers of moles of ions per second which traverse a 1 cm^2 section are given:

– for cations, by: $u_{0+}\nu_+C/1000$;

– for anions, by: $u_{0-}\nu_-C/1000$.

Thus, the amount of electricity passing across that surface each second is:

$$\chi_0 = \left(u_{0+}\nu_+ + u_{0-}\nu_-\right)\frac{C\mathfrak{F}}{1000} \qquad [1.32]$$

In view of relations [1.22] and [1.24], we obtain the following for the limiting equivalent conductivity:

$$\Lambda_0 = \left(u_{0+}\nu_+ + u_{0-}\nu_-\right)\mathfrak{F} \qquad [1.33]$$

This limiting equivalent conductivity is the sum of two contributions:

– one contribution made by the cations:

$$\lambda_{0+} = u_{0+}\mathfrak{F} \qquad [1.34]$$

– one contribution from the anions:

$$\lambda_{0-} = u_{0-}\mathfrak{F} \qquad [1.35]$$

The values λ_{0+} and λ_{0-} are called the *limiting equivalent ionic conductivities*. These two contributions are independent of one another, because the limiting mobilities are values which are intrinsic to each individual ion. It follows that the limiting equivalent conductivity is the sum:

$$\Lambda_0 = \nu_+ \lambda_{0+} + \nu_- \lambda_{0-} \qquad [1.36]$$

This is known as Kohlrausch's law, which was discovered through experimentation.

Based on tables showing the limiting mobilities, or the limiting equivalent ionic conductivities λ_{0+} and λ_{0-}, it is possible to calculate the limiting equivalent conductivity for a given fully-dissociated electrolyte. Table 1.1 gives an extract of such a table.

Cation	λ_{0+} ($\Omega^{-1}.mol^{-1}.cm^2$)	Anions	λ_{0-} ($\Omega^{-1}.mol^{-1}.cm^2$)
H^+	349.60	OH^-	199.1
Li^+	38.69	Cl^-	76.34
Na^+	50.11	Br^-	78.40
K^+	73.50	I^-	76.80
Mg^{++}	106.12	SO_4^{--}	159.60
Ca^{++}	119.00	NO_3^-	71.40
Ba^{++}	127.28	$CH_3CO_2^-$	40.90

Table 1.1. *Limiting equivalent ionic conductivities of a number of ions*

To establish the individual values of the limiting ionic mobilities (Table 1.1), we combine the use of relation [1.36] and the measurement of the limiting conductivity of a strong electrolyte with a measurement of the transport numbers in that same electrolyte. We then use the following relation, which is easy to prove:

$$\frac{\lambda_{0+}}{\lambda_{0-}} = \frac{\tau_+}{\tau_-}\frac{\nu_-}{\nu_+} \qquad [1.37]$$

NOTE.– The only hypotheses made in this section are complete dissociation of the electrolyte and a concentration which tends toward zero, so relations [1.33] and [1.36] apply both the completely-dissociated strong electrolytes and to weak electrolytes because we know that, at infinite dilution, the dissociation coefficient tends toward 1, and that the weak electrolyte tends toward complete dissociation and becomes a strong electrolyte.

1.4.5. *Apparent dissociation coefficient and equivalent conductivity*

We use the term *apparent dissociation coefficient* α_a to denote the dissociation coefficient of a weak electrolyte whose ions have the same mobility at a given concentration C as at zero dilution. The number of moles of ions per second crossing a 1cm^2 cross-section under the influence of the unit field is:

– for cations: $u_{0+}\nu_+ C\alpha_a / 1000$;

– for anions: $u_{0-}\nu_- C\alpha_a / 1000$.

From this, we deduce a conductivity as follows:

$$\chi = \left(u_{0+}\nu_+ + u_{0-}\nu_-\right)\frac{C\alpha_a\mathfrak{F}}{1000} \qquad [1.38]$$

and an equivalent conductivity at concentration C:

$$\Lambda_0 = \left(u_{0+}\nu_+ + u_{0-}\nu_-\right)\alpha_a\mathfrak{F} \qquad [1.39]$$

By comparing relations [1.33] and [1.39], we can deduce the apparent dissociation coefficient:

$$\alpha_a = \frac{\Lambda}{\Lambda_0} \qquad [1.40]$$

Whilst we have hitherto considered the mobilities to be independent of the concentration, it has long been held that this apparent degree of dissociation is the true degree of dissociation at concentration C. Thus, the variation of the equivalent conductivity values was attributed to dissociation alone.

1.4.6. *Variations of equivalent conductivities with the concentrations*

We shall now take a look at the variations of the mobility values with concentration, using expressions [1.30a] and [1.30b]. In turn, we examine the cases of completely-dissociated strong electrolytes and of weak electrolytes exhibiting a true dissociation coefficient α at concentration C.

1.4.6.1. *Case of strong electrolytes*

Let us begin by considering a strong electrolyte at concentration C, completely dissociated. The ionic mobilities of its ions at that concentration are u_+ and u_-, and its equivalent conductivity is Λ. We can employ exactly the same reasoning as in section 1.2.3, giving us the following relation, which is valid for the concentration C:

$$\Lambda = (u_+\nu_+ + u_-\nu_-)\mathfrak{F} \quad [1.41]$$

Thus, by defining ionic mobilities at concentration C by $\lambda_+ = u_+\mathfrak{F}$ and $\lambda_- = u_-\mathfrak{F}$, we obtain the relation:

$$\Lambda = \nu_+\lambda_+ + \nu_-\lambda_- \quad [1.42]$$

Kohlrausch's law therefore remain valid at the concentration C for completely dissociated electrolytes.

If we feed expressions [1.32] back into relation [1.41], and recall that in the case of a single electrolyte, the ionic strength is proportional to the concentration of that electrolyte C, we can use the Debye, Hückel and Onsager law to show that, within the limits of concentrations of validity of the Debye–Hückel solution limiting law, the conductivity of the strong (completely dissociated) electrolyte is given by:

$$\Lambda = \Lambda_0 - (B_1\Lambda_0 + B_2)\sqrt{C} \quad [1.43]$$

In this relation, the coefficients B_1 and B_2 are constants which depend only on the solvent (through its dielectric constant and its viscosity), the charges of the ions and the temperature.

For example, for a 1-1 electrolyte – i.e. an electrolyte for which we have $\nu_+ = \nu_- = z_+ = z_- = 1$, those coefficients are given in water by Table 1.2 at two temperatures.

T (°C)	B_1	B_2
18°C	0.224	50.5
20°C	0.229	59.5

Table 1.2. *Coefficients B_1 and B_2 from relation [1.43] in water at two temperatures*

NOTE.– Relation [1.43] can also be written in the form:

$$\frac{\Lambda}{\Lambda_0} = \varphi = 1 - \left(B_1 + \frac{B_2}{\Lambda_0} \right) \sqrt{C} \qquad [1.44]$$

In this expression, which is only valid for a strong electrolyte, the coefficient ϕ, which is called the conductivity coefficient, no longer represents an apparent dissociation coefficient. It merely enables us to compare the equivalent conductivity at concentration C to the limiting equivalent conductivity of a completely dissociated electrolyte, this coefficient actually quantifies the effect of dilution.

From relation [1.43], we see that it is possible to determine the limiting equivalent conductivity of a strong electrolyte by extrapolation to zero concentration of the straight line showing the equivalent conductivity as a function of the square root of the concentration. Figure 1.1 shows the example, in curve *a*, of such a determination for potassium chloride solutions.

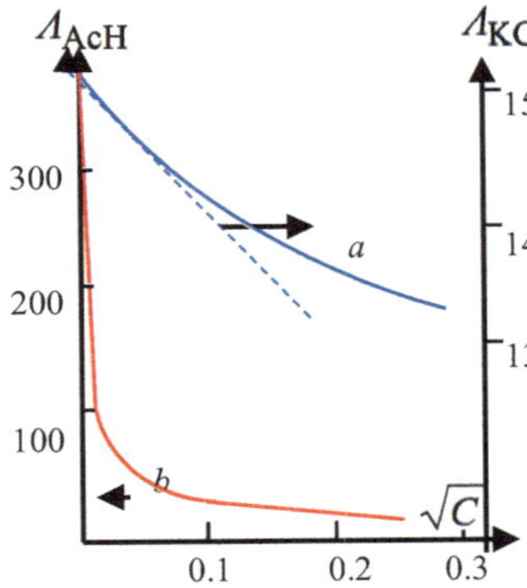

Figure 1.1. *Equivalent conductivity a) of potassium chloride – b) of acetic acid as a function of the square root of the concentration*

1.4.6.2. *Case of weak electrolytes*

In the case of weak electrolytes that are not completely dissociated, we can no longer apply relation [1.44]. To solve the problem we shall use the MacInnes and Shedlovsky method. Those authors introduce a completely-dissociated fictitious electrolyte containing the same ions as the weak electrolyte under study. This electrolyte would be the effective concentration C_e which is the same as the concentration of the ions in the real electrolyte, so if α is the dissociation coefficient of our electrolyte, the effective concentration will be:

$$C_e = \alpha C \qquad [1.45]$$

The equivalent conductivity of that strong electrolyte at the concentration C_e is Λ_e. To calculate it, as we are dealing with a strong electrolyte, we can apply relation [1.44], which gives us:

$$\frac{\Lambda_e}{\Lambda_0} = \varphi_e = 1 - \left(B_1 + \frac{B_2}{\Lambda_0} \right) \sqrt{\alpha C} \qquad [1.46]$$

The ratio Λ/Λ_e for our weak electrolyte will only take account of the effect dissociation, as the effect of dilution is canceled out because the two electrolytes with equivalent conductivity values Λ and Λ_e are at the same concentration.

Thus, for the weak electrolyte under study here with equivalent conductivity Λ, we can write the following, if α is the dissociation coefficient:

$$\frac{\Lambda}{\Lambda_0} = \frac{\Lambda}{\Lambda_e}\frac{\Lambda_e}{\Lambda_0} = \alpha\varphi_e = \alpha\left[1 - \left(B_1 + \frac{B_2}{\Lambda_0} \right) \sqrt{\alpha C} \right] \qquad [1.47]$$

This relation can also be written thus:

$$\Lambda = \Lambda_0 \alpha - \left(A_1 \Lambda_0 + A_2 \right) \alpha^{3/2} \sqrt{C} \qquad [1.48]$$

We can see that the function [1.48] giving Λ as a function of $\sqrt{C}$ is, in fact, more complex than it appears, because the dissociation coefficient is itself a function of C. Figure 1.1 shows the curve b which represents Λ for ethanoic acid.

It stems from the shape of the curve b that Λ_0 cannot be determined by extrapolation. We can, however, determine this value using Kohlrausch's law – e.g. in the case of ethanoic acid AcH, we can write:

$$\Lambda_0(\mathrm{HAc}) = \lambda_0(\mathrm{H}^+) + \lambda_0(\mathrm{Ac}^-) \qquad [1.49]$$

This sum can be broken down as follows:

$$\begin{aligned}\Lambda_0(\mathrm{HAc}) = \lambda_0(\mathrm{H}^+) + \lambda_0(\mathrm{C}l^-) + \left(\lambda_0(\mathrm{N}a^+) + \lambda_0(\mathrm{Ac}^-)\right) \\ -\left(\lambda_0(\mathrm{N}a^+) + \lambda_0(\mathrm{C}l^-)\right)\end{aligned} \qquad [1.50]$$

In the pairs of terms, we recognize the limiting equivalent conductivities of the three compounds HCl, NaAc and NaCl, so we have:

$$\Lambda_0(\mathrm{HAc}) = \Lambda_0(\mathrm{HCl}) + \Lambda_0(\mathrm{NaAc}) - \Lambda_0(\mathrm{NaCl}) \qquad [1.51]$$

The three necessary values $\Lambda_0(\mathrm{HCl})$, $\Lambda_0(\mathrm{NaAc})$ and $\Lambda_0(\mathrm{NaCl})$, are easy to determine separately on the basis of the pure electrolytes, using the extrapolation method, described in section 1.4.6.2, because HCl, NaAc and NaCl are strong electrolytes and therefore obey relation [1.43].

1.5. Determination of the dissociation coefficient

We shall now describe two generally methods which can theoretically be used to determine the dissociation coefficients. One is based on cryometry, and the other on electrical conductivity. In Chapter 7 (section 7.7.2) we shall see a method for determining the dissociation constant of a weak electrolyte on the basis of the voltage of a battery.

1.5.1. *Determination of the dissociation coefficient by the cryometric method*

The first method is thermodynamic, because the ionic dissociation has a direct influence on Raoult's laws of ebullioscopy and cryoscopy. Indeed, in the case of cryoscopy, the drop in the melting point of the solvent (index 0) is given by the relation:

$$\Delta T_{0(F)} = \frac{RT_{0(F)}^2}{\Delta_F h_0^0} x_1 \qquad [1.52]$$

If the solution now contains not one but several solutes i, we can easily show that the drop of the melting point becomes:

$$\Delta T_{0(F)} = \frac{RT_{0(F)}^2}{\Delta_F h_0^0} \sum_i x_i \qquad [1.53]$$

Suppose that the molecule dissociating gives rise to ν ions, and α is the degree of ionization. If we initially consider n_s moles of non-dissociated solute, the total amount of solute n_t after partial dissociation is:

$$n_t = n_s(1-\alpha) + \alpha n_s \nu = n_s\left[1+(\nu-1)\alpha\right] \qquad [1.54]$$

The sum of the molar fractions of the solutes is therefore:

$$\sum_i x_i = \frac{n_s\left[1+(\nu-1)\alpha\right]}{n_0 + n_s\left[1+(\nu-1)\alpha\right]} \qquad [1.55]$$

If the solution is sufficiently dilute, the second term in the denominator in the previous fraction is small in relation to n_0 (amount of solvent) and thus the sum of the molar fractions is approximately:

$$\sum_i x_i = x_s\left[1+(\nu-1)\alpha\right] \qquad [1.56]$$

We see the emergence of a factor j defined by:

$$j = \left[1+(\nu-1)\alpha\right] \qquad [1.57]$$

This factor j is called the *van 't Hoff factor*. It can be measured by comparing the cryoscopic drop measured with the electrolyte solution with the calculated value which would be obtained without dissociation – i.e. for $j = 1$.

We supposed, when using relation [1.53], that the solution was sufficiently dilute, meaning that it exhibited ideal behavior. If this is not the case, and we know that this is more common with ionic solutions than for molecular solutions, we must take account of the mean activity coefficient of the ions $\gamma_{\pm}$.

In practice, the differences between the theoretical value and the experimental value are too low for the cryometric method to be able to be used to accurately determine a degree of ionization. Cryometry in a molten-salt medium is able to deliver a more precise result.

1.5.2. Determination of the dissociation coefficient on the basis of the conductivity values

To determine the degree of ionization, it is preferable to use a dynamic method based on measuring the equivalent conductivity of the electrolytical solution at the desired concentration and calculating the limiting equivalent conductivity. This method is limited to a concentration domain below 0.02 moles per liter, which is the domain of validity of the Debye–Hückel theory for modeling strong electrolytes.

To determine the dissociation coefficient of an electrolyte at a concentration C, we measure its equivalent conductivity and calculate its limiting equivalent conductivity by an addition reaction similar to expression [1.50]. We then use expression [1.48] to calculate α.

In practice, to determine α on the basis of relation [1.48], which cannot be expressed in terms of α, we carry out a succession of approximations.

We determine a first approximate value for α: α_1 by making $\alpha = 1$, so:

$$\alpha_1 = \frac{\Lambda}{\Lambda_0} \qquad [1.58]$$

We feed this value back into the expression of α, which enables us to calculate a second approximation α_2 by using the relation:

$$\alpha_2 = \frac{\Lambda}{\Lambda_0} \frac{1}{1 - (B_1 \Lambda_0 + B_2)\sqrt{\alpha_1 C}} \quad [1.59]$$

The operation is repeated as many times as necessary. In practice, usually, the second calculated value of α is sufficient and therefore $\alpha = \alpha_2$.

It is the application of relation [1.59], using the Debyean terms B_1 and B_2, which limits the method to solutions containing concentrations lower than 0.02 moles/l, i.e. within the bounds of application of Debye's model.

1.6. Determination of the number of ions produced by dissociation

We shall now describe two experimental methods that can be used to determine the number of ions produced by the dissociation of an electrolyte. These methods, which can be used for strong (i.e. completely dissociated) electrolytes, are again based on measurements of the conductivity and cryometric measurements.

1.6.1. *Use of limiting molar conductivity*

We have seen (section 1.4.3) that the ionic mobilities do not vary within a very large range (between 2×10^{-4} and 10^{-3} $cm^2s^{-1}V^{-1}$, with the exception of the ions H^+ (3×10^{-3} $cm^2s^{-1}V^{-1}$) and OH^- (2×10^{-3} $cm^2s^{-1}V^{-1}$). The same is true of the limiting equivalent conductivities. Consequently, the limiting molar conductivity will increase with the number of ions provided by the molecule. Table 2.1 shows the ranges of the molar conductivities depending on the number of ions supplied by the decomposition of the molecule.

Obviously, this method lacks precision and can only be applicable for dissociated species. Thus, if we consider the three complexes in platinum – $[Pt(NH_3)_4Cl_2]Cl_2$, $[Pt(NH_3)_3Cl_3]Cl$ and $[Pt(NH_3)_4Cl_4]$ – upon dissociation they can only furnish chlorine ions and complex ions. For these three complexes, we find the limiting molar conductivities 228, 97 and 1. The first value belongs to the range 200–300 and thus there will by three ions (one complex ion and two chlorine ions). The second value belongs to the range 100–200, so there are two ions (one complex ion and one chlorine ion). Finally, for the last complex, the value belongs to the range < 100, so it gives no ions at all.

Number of ions	Molar conductivity
2	100–150
3	200–300
4	400–450
5	> 500

Table 1.3. *Molar conductivity values and number of ions*

1.6.2. *Use of cryometry*

The method presented here applies to completely dissociated (i.e. strong) electrolytes. We employ the same reasoning as in section 1.5.1 but adapt it to the decomposition of an electrolyte, which ionizes to produce n ions. By virtue of relations [1.53] and [1.56], we can write the following for the variation of the melting point of the solvent:

$$\Delta T_{0(F)} = \frac{\mathrm{R} T_{0(F)}^{2}}{\Delta_F h_0^0} x_s \left[1 + (n - 1)\alpha\right] \quad [1.60]$$

Let us introduce the quantity q_s (number of moles) of solute for 1000 g of solvent. It is easy to show that, for a dilute solution ($x_s \approx 1$), this quantity is given as a function of the molar fraction of the solute and the molar mass of the solvent by the relation:

$$q_s = \frac{100 x_s}{M_0} \quad [1.61]$$

Thus, the variation in melting point is written as:

$$\Delta T_{0(F)} = \frac{\mathrm{R} T_{0(F)}^{2} M_0}{1000 \Delta_F h_0^0} q_s \left[1 + (n - 1)\alpha\right] \quad [1.62]$$

We see the emergence of the cryoscopic constant of the solvent, which is defined by:

$$K_{0(Cryo)} = \frac{\mathrm{R}T_{0(F)}^{2}M_{0}}{1000\Delta_{F}h_{0}^{0}} \qquad [1.63]$$

The variation in temperature of the solvent therefore becomes:

$$\Delta T_{0(F)} = K_{0(Cryo)}q_s\left[1+(n-1)\alpha\right] \qquad [1.64]$$

If there were no ionization of the electrolyte, that variation in the melting point would be, where ($n = 0$):

$$\Delta T_{0(F)} = K_{0(Cryo)}q_s \qquad [1.65]$$

Let us examine the limit of the *molecular drop* $\frac{\Delta T_{0(F)}}{q_s}$, when q_s tends toward 0. In these conditions, when dilution tends toward infinity, we know that the fraction of dissociation of the complex tends toward 1, so we have:

$$\lim_{q_s \to 0}\frac{\Delta T_{0(F)}}{q_s} = K_{0(Cryo)}n \qquad [1.66]$$

If the electrolyte were not dissociated, that ratio would be constant:

$$\frac{\Delta T_{0(F)}}{q_s} = K_{0(Cryo)} \qquad [1.67]$$

Figure 1.2 shows the variations in the molecular drop with the quantity of complex dissolved.

In the case of non-dissociation, the curve is not a horizontal because the variations of the activity coefficients with dilution come into play.

In the case of dissociation, the ordinate at the origin is used to calculate the coordination number if we know the cryoscopic constant of the solvent, which can be calculated.

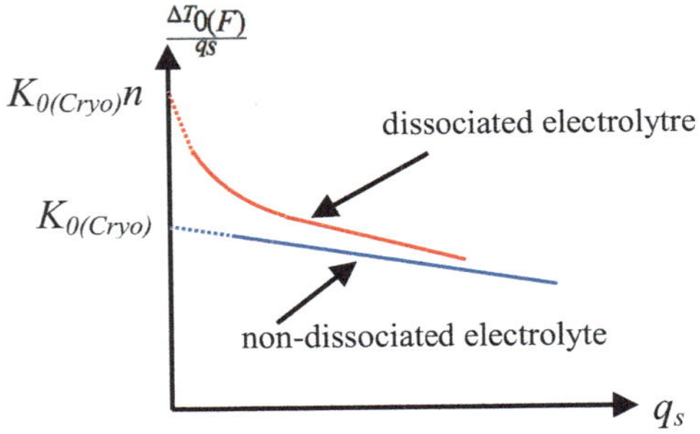

Figure 1.2. *Variation of the molecular drop with the quantity of complex dissolved*

In practice, the variation of the activity coefficients, which is very significant at high dilution, renders the above extrapolation very difficult. It is preferable to operate in conditions such that the activity coefficients practically do not vary. With this in mind, we carry out cryoscopy in a molten-salt medium.

Numerous saline hydrates behave like cryoscopic solvents when we dissolve a substance in the molten hydrate. We observe a cryoscopic drop of the melting point. We also see a cryoscopic drop in the temperature of the hydrate-to-anhydrous transformation and the temperature of a eutectic, e.g. between a salt and ice.

In this case, the activity coefficients of the ions practically no longer vary with dilution. Indeed, we know that these activity coefficients vary not with the composition of an ionic solution but with the ionic strength of the solvent. Molten salts are composed of a large quantity of ions so that the addition of a solute does not alter the ionic strength of the medium.

In these conditions, the molecular drop varies little with the quantity of dissolved salt, and therefore it is easy to extrapolate to the situation of zero dissolved quantity. In addition, if the solvent contains a common ion with the solute, that ion is inactive from the point of view of cryoscopy, because its quantity practically does not vary when the complex is dissociated.

The method applies both to the dissociation of complexes and that of the ions.

For example, with cryoscopies at the transition temperature between sodium sulfate and sodium sulfate decahydrate, it has been possible to measure the following values of n:

– n = 1; for sodium chloride NaCl (the common sodium ion has no influence), and for magnesium sulfate $MgSO_4$ (with a common sulfate ion). The true value of the coordination number is 2;

– n = 2, for potassium nitrate KNO_3 and potassium sulfate K_2SO_4 (common sulfate ion). The true value of the coordination number is 3;

– n = 3, for magnesium chloride $MgCl_2$.

1.7. Thermodynamic values relative to the ions

In order to study the ionic medium, we need to know the enthalpies and entropies of the reactions, and therefore the enthalpies and entropies of formation of the ions. This is the topic of our discussion in this section.

Note, firstly, that it is not possible to deduce the values we seek from the known values for the atoms and molecules because, in view of the necessity to respect the electrical neutrality of the balance equations, it is impossible to design reactions which involve only the ions of one species.

1.7.1. *The standard molar Gibbs energy of formation of an ion*

Consider, for instance, the reaction of formation of a pair of dissolved ions:

$$A_{pure} + B_{pure} + \text{solvent} = A^+ \text{ (dissolved)} + B^- \text{ (dissolved)} \qquad [1R.5]$$

The Gibbs energy associated with that reaction is given by:

$$\begin{aligned} \Delta_5 G_T &= \mu_T^0(A^+) - \mu_T^0(B^-) + RT\ln\frac{|A^+|}{|B^-|} \\ &= \mu_T^0(A^+) - \mu_T^0(B^-) + RT\ln\frac{[A^+]}{[B^-]} + RT\ln(\gamma_\pm)^{(\nu_+ + \nu_-)} \end{aligned} \qquad [1.68]$$

The standard chemical potentials of the ions cannot be found individually. At most, by placing ourselves in conditions such that the mean activities of the ions are equal to 1, and considering the Gibbs energy associated with the dilution, we can determine the difference between the two standard potentials $\mu_T^0(\mathrm{A}^+) - \mu_T^0(\mathrm{B}^-)$.

To find an individual value, it is necessary, by convention, to attribute a particular value to the standard chemical potential of an ion, and then, starting with that chosen value, little by little, calculate a complete scale of the standard potentials of the ions.

To choose the base, we consider the electrochemical reaction associated with the neutralization of the proton, which is written thus:

$$\mathrm{H}^+ + \mathrm{e}^- = 1/2\mathrm{H}_2 \qquad [1R.6]$$

The standard Gibbs energy associated with this reaction is:

$$\Delta_6 g_{298}^0 = \frac{1}{2}\mu_{298}^0(\mathrm{H}_2) - \mu_{298}^0(\mathrm{H}^+) \qquad [1.69]$$

By convention, this standard Gibbs energy is chosen as being equal to 0 at 25°C. This is known as *Latimer's first convention*. Thus:

$$\Delta_6 g_{298}^0 = 0 \qquad [1.70]$$

By combining relations [1.69] and [1.70], we immediately obtain:

$$\mu_{298}^0(\mathrm{H}^+) = \frac{1}{2}\mu_{298}^0(\mathrm{H}_2) \qquad [1.71]$$

As the standard enthalpy of formation of the hydrogen molecule is zero at 25°C, we can deduce that:

$$\mu_{298}^0(\mathrm{H}^+) = -\frac{298}{2} s_{298}^0(\mathrm{H}_2) \qquad [1.72]$$

Thus, the standard chemical potential of the hydrogen ion is directly linked to the standard entropy of the dihydrogen molecule.

At a temperature other than 25°C, relation [1.71] still holds true, but the standard enthalpy of formation of the hydrogen molecule is no longer zero.

1.7.2. *Standard enthalpy of formation of ions*

By convention, and for the same reasons as for the Gibbs energy, *Latimer's second convention* is to posit that the standard enthalpy of formation of H^+ ions is 0 at any given temperature.

$$h_T^0(H^+) = 0 \qquad [1.73]$$

This convention is the same as that of neutral molecules of simple substances, which is normal because relation [1.71] can be transposed to apply to enthalpies.

1.7.3. *Absolute standard molar entropy of an ion*

In view of convention [1.73], relation [1.72] becomes:

$$s_T^0(H^+) = \frac{1}{2} s_T^0(H_2) \qquad [1.74]$$

Hence, at 25°C:

$$s_{298}^0(H^+) = \frac{1}{2} s_{298}^0(H_2) = 65\ 342\ \text{J.mole}^{-1}\text{K}^{-1} \qquad [1.75]$$

Note that the entropy of formation of the H^+ ions is not zero, contrary to another convention which Latimer had chosen and which thus proved to be inconsistent with the first two.

In this chapter, we have examined the behavior of an electrolyte placed in water. The study would have been performed in the same way if, instead of water, we had used another solvent, provided that, like water, that solvent has a high dielectric constant. In the case of a solvent with a lower dielectric constant, we see that the solution behaves as though the number of ions were less. The conductivity decreases. Not only are the coefficients B_1 and B_2 from relation [1.41] different, but it is as if the concentration too were lower – even in the case of a strong electrolyte in the solvent at hand. In

Chapter 3, when we study acids, we shall see the role of the dielectric constant of the solvent on the dissociation of an acid. Then, we introduce the ion pair model.

1.7.4. Determination of the mean activity of a weak electrolyte on the basis of the dissociation equilibrium

Suppose we know the dissociation constant of the weak electrolyte AH and its dissociation coefficient α at the chosen concentration C. The dissociation constant is written as:

$$K = \frac{\alpha^2 C}{1-\alpha} \frac{(\gamma_{\pm})^2}{\gamma_{AH}} \qquad [1.76]$$

Adopting the hypothesis of a solution sufficiently dilute in terms of electrolyte AH, we know that as the concentration decreases, neutral molecules approach an ideal state much more quickly than do ions. Thus, we shall posit $\gamma_{\mathrm{AH}} = 1$. Immediately, we find;

$$\gamma_{\pm} = \sqrt{\frac{K(1-\alpha)}{\alpha^2 C}} \qquad [1.77]$$

We shall see other methods for measuring the mean activity of an electrolyte in Chapter **7**.

2

Solvents and Solvation

2.1. Solvents

One of the most commonly used methods to obtain an ionic solution is to mix a solvent with a solid compound exhibiting marked electrovalence.

Solvents can be classified on the basis of how easily they can separate ions. The classification system is based on the value of the dielectric constant. Indeed, the force of attraction between ions of opposite signs is inversely proportional to the dielectric constant; hence, the higher the value of that constant, the weaker the inter-ionic forces, and therefore the stronger the solvent's ionizing power.

There may be significant differences between the values of the dielectric constants of solvents. Table 2.1 gives an example of the dielectric constants D of a number of ionizing solvents.

Solvents such as carbon tetrachloride (D = 2.24), ethyl carbonate (D = 2.8) or benzene (D = 2.27) are considered to be inert solvents without an ionizing power.

Solvent	D
Cyanhydric acid	114
Formaldehyde	109
Water	78
Ethanol	25
Liquid ammonia	20
1,4-Dioxane	9.5
Acetic acid	6

Table 2.1. *Values of some dielectric constants of ionizing solvents*

Another classification system divides solvents into three categories depending on the nature of their bond:

– molecular solvents, which are essentially composed of electrically-neutral molecules. The molecules are linked to one another by hydrogen bonds or van der Waals forces, and the intramolecular bonds are covalent. They are liquids whose dielectric constant varies in an extended domain ranging from 2 for benzene to 165 for N-methylacetamide. They do not conduct electrical current well because they have a low degree of ionization. This is the family to which water, liquid ammonia and numerous organic and inorganic compounds belong;

– ionic solvents, composed of anions and cations. The bonds in these solvents are strongly ionic, which explains why, in order to use liquids as a solvent, it is necessary to raise them to a high temperature. Essentially, ionic solvents are molten salts. To lower the working temperature, we can use eutectic mixtures, such as cryolith – a mixture of sodium fluoride and aluminum fluoride ($3FNa + F_3Al$), which is used for the electrolysis of aluminum;

– metallic solvents or molten metals. Cohesion is assured by metallic bonds. These are excellent electrical conductors. This category includes liquids at very different temperatures, including the ordinary temperature depending on the melting point of the metal or the eutectic used.

Finally, a third useful distinguishing factor between solvents is their ability (or lack thereof) to exchange protons with another molecule. Thus, we distinguish *protonic* solvents from *aprotic* solvents, which have no capacity for such proton exchanges. For instance, sulfur dioxide can be used as a solvent, and is aprotic.

Protonic solvents can, themselves, be classified into three categories in terms of their acid/base properties. We distinguish:

– acidic or *protogenic* solvents which act solely as proton donors. This category includes sulfuric acid, acetic (or ethanoic) acid), formic (or methanoic) acid, acetone (or propanone);

– basic or *protophilic* solvents which act solely as proton acceptors. In this category, we can cite pyridine, liquid ammonia, amines and 1,4-dioxane;

– *amphoteric* or *amphiprotic* solvents, which are capable of acting as a proton donor and/or acceptor. This category includes water, alcohols and acetonitrile.

2.2. Solvation and structure of the solvated ion

As we saw in section 1.1.1, solvation is the fixation of a certain number of molecules of solvent onto the ions. This solvation becomes important in two fields:

– the study of the mobility of ions and therefore of conductivity, because the motions of the ions are completely different depending on whether or not they carry with them a certain number of molecules of solvent;

– the study of the influence of the solvent on reactions in solution. Obviously, the ions' behaviors are dependent on their immediate environment.

As is the case with pure liquids, it is possible to study the spatial distribution of species around an ion using the radial distribution function[1], which can be determined by X-ray diffraction, or better still, neutron diffraction.

Remember that the radial distribution $g_{i,j}(r)$ is the probability of finding an atom i at the distance r from an atom j. This probability is also known as the *pair distribution*.

Figure 2.1 shows the pair distribution functions obtained for the Cl-O and Cl-H pairs in an aqueous solution of sodium chloride. We can plainly see, on the two curves, the presence of a first, highly-pronounced peak which indicates significant structuring of water molecules in the immediate vicinity of the chlorine ion. The next maximum is much attenuated, and the third is barely visible.

1 On this subject, readers could usefully consult the first chapter of Volume 2 of this collection, dedicated to liquid phases [SOU 15b].

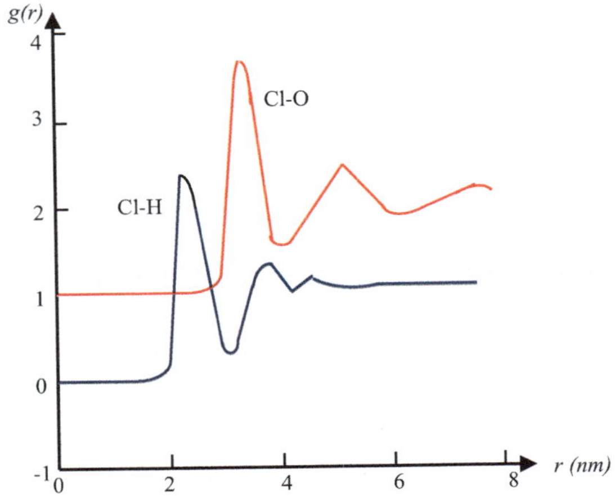

Figure 2.1. *Pair distribution functions for Cl-H and Cl-O in salt water (from [IMP 86])*

This means that in the immediate vicinity of the chlorine ion, the water molecules are very clearly localized, forming what is known as the first solvation layer (Figure 2.2). Then comes a fairly well-structured zone which is known as the buffer zone, in which the molecules exhibit a certain degree of order.

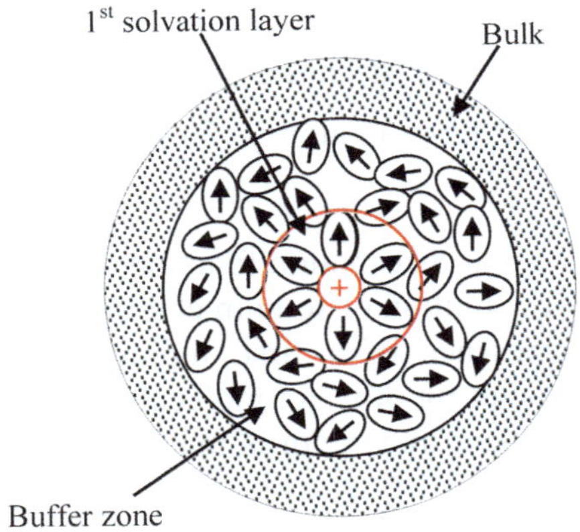

Figure 2.2. *Arrangement of dipolar molecules of solvent around the solvated ion*

Finally, the rest of the solvent constitutes the bulk. In this zone, the central ion no longer exerts any influence whatsoever on the molecules of solvent.

On the basis of the distribution function, it is possible to calculate the number of atoms constituting the first solvation layer, using the relation:

$$z = 4\pi\rho \int_0^{r_{\min}} r^2 g(r)\,\mathrm{d}r \qquad [2.1]$$

In this relation, we decide that the immediate neighbors are those solvent atoms at distances between 0 and r_{min} from the ion. This latter value is the abscissa of the minimum which comes after the first maximum (Figure 2.1) of $g(r)$.

This number is usually 6 in water for most ions such as halogen or transition metal ions. The value is a little less with the smallest ions. For example, it is 4 in the case of Li^+ ions, and 4 to 6 for the Na^+ and K^+ ions.

By taking measurements – notably by magnetic resonance relaxometry – it is possible to measure the mean time of stay of the molecules of solvent in the different layers formed around the ion. We note that the ions in the first layer exchange only very slowly with molecules outside of that layer, unlike the molecules in the buffer zone, which are much more "mobile". Beyond the second layer, the times of stay are the same as they are in pure water. All these exchange times vary with the size and charge of the central ion.

2.3. Thermodynamics of solvation

Ultimately, solvation is the result of a reaction which is written as:

$$\text{A (gas)} \rightarrow \text{A (dissolved)} \qquad [2R.1]$$

This reaction is possible for neutral molecules or atoms, such as a rare gas, just the same as it is with an ion. For example, in the case of the sodium ion, this reaction is written:

$$Na^+\text{(gas)} \rightarrow Na^+ \text{(dissolved)} \qquad [2R.2]$$

It expresses the passage of the sodium ion from the gaseous state to the solvated state in the solution.

Solvation where water is the solvent has the particular name of *hydration*.

2.3.1. *Thermodynamic values of solvation*

The Gibbs energy of solvation and the enthalpy of solvation of an atom, a molecule or an ion are, respectively, the Gibbs energy and the enthalpy associated with the reaction of solvation [2R.1].

For a neutral species, whether solid, liquid or gaseous, we can measure the enthalpy of solvation using a calorimeter, breaking a vial containing the product in question, which is immersed in a large quantity of solvent.

The Gibbs energy of solvation can be estimated by calculating the work required to form a cavity of surface area equal to that of the molecule. For example, we estimate the Gibbs energies of solvation of rare gases as the amount of work necessary to create a spherical cavity whose radius is equal to that of the atom of the rare gas. These calculations are performed when we know the surface tensions[2] of the solvents.

We know that for rare gases, for example, the Gibbs energy of solvation is slightly positive (around 25 kJ/mole in water, and 20 kJ/mol in benzene). Therefore, it is more favorable for the gases to exist in the gaseous state than to be present as dissolved gases. Thus, these gases are not greatly soluble in water.

In the case of ions, we distinguish between two families of thermodynamic values of solvation: the *absolute* values and the *relative* values.

Indeed, given the impossibility of solvating a species of ion in isolation, we have, as we did for the Gibbs energies of formation, expressed the Gibbs energies of solvation on a scale where the standard Gibbs energy of solvation of the proton is taken as being equal to 0. Thus, we posit that at all temperatures, for the reaction:

$$H^+ \text{ (gas)} = H^+ \text{ (dissolved)} \qquad [2R.3]$$

2 This type of calculation will be used in Volume 7 of this series, which is dedicated to surface phenomena.

$$\Delta_{solv} G^0_{\mathrm{H}^+} = 0 \qquad [2.2]$$

We do likewise for the standard enthalpy:

$$\Delta_{solv} H^0_{\mathrm{H}^+} = 0 \qquad [2.3]$$

Obviously, the conclusion to draw from this is that we have:

$$\Delta_{solv} S^0_{\mathrm{H}^+} = 0 \qquad [2.4]$$

The *relative standard Gibbs energy of solvation* of a cation is, in fact, the standard Gibbs energy of the reaction:

$$\mathrm{M}^+ \text{ (gas)} + \nu \mathrm{H}^+ \text{ (dissolved)} = \mathrm{M}^+ \text{ (dissolved)} + \nu \mathrm{H}^+ \text{ (gas)} \qquad [2R.4]$$

Thus:

$$\Delta G^0_M = \Delta G^0_4 - \Delta G^0_3 \qquad [2.5]$$

Using sophisticated computational methods, it has been possible to find absolute values for the Gibbs energy and enthalpy of solvation of a proton by water. The following values can be taken as accurate.

$$\Delta_{solv} G^0_{\mathrm{H}^+} = -1090 \text{ kJ/mol} \quad \text{and} \quad \Delta_{solv} H^0_{\mathrm{H}^+} = -1130 \text{ kJ/mol}$$

Based on these values, it is obviously easy to calculate, little by little, the *standard absolute enthalpy and Gibbs energy values of hydration* of the ions. Note that for highly-soluble salts, the values obtained are strongly negative, demonstrating the stability of ions solvated by water in comparison to gaseous ions. For instance, for the Na^+ ion, we obtain an approximate value of -205 kJ/mol.

2.3.2. *Gibbs energy of salvation – Born's model*

In 1920, Born offered up a model to calculate the absolute Gibbs energy of solvation of an ion in a solvent. In fact, this model calculates only the electrical contribution to that solvation, only taking account of the interactions between the ion and the electrical field created by that ion's charge. In reality, the Gibbs energy of solvation of an ion is the sum of two terms: the electrical term, which we have just defined, and a term due to the short-distance interactions, by way of van der Waals forces, between the ion A_i and the molecules of solvent which are solvating it. Thus, in a solvent S, we can write:

$$\Delta_{sol} G_i^{0(S)} = \Delta_{elec} G_i^{0(S)} - \Delta_{vdW} G_i^{0(S)} \quad [2.6]$$

To explain Born's model, which therefore applies only to the electric part, we consider the cycle shown in Figure 2.3.

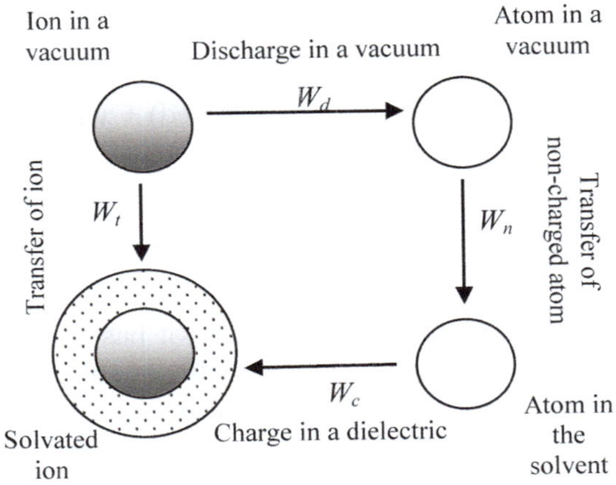

Figure 2.3. *Born's model of solvation*

Consider a spherical ion with radius r_i. The energy W_t corresponds to the Gibbs energy of dissociation, because it is the energy accompanying reaction [2R.2].

In this cycle, the energy W_n can be considered to be the Gibbs energy of solvation of a spherical, non-charged atom. In other words, it is essentially that of the rare gas with the same radius.

W_d and W_c are respectively the opposite of the charge energy in a vacuum, and the charge energy in the dielectric with constant D of the spherical ion.

We have displayed these different energies on the cycle in Figure 2.3. From this cycle, we see that we can write:

$$W_t - W_n = W_d + W_c \qquad [2.7]$$

This difference $W_t - W_n$ represents the energy of interaction between an ion and the solvent, because it represents the difference between the Gibbs energy of solvation of an ion and that of a neutral atom or molecule of the same size. It is the electrical contribution to the Gibbs energy of solvation.

Let $\mu_i^{0(\mathrm{Gas})}$ denote the standard chemical potential (P = 1 bar) of the ion A_i in the gaseous phase. $\mu_{GR}^{0(\mathrm{Gas})}$ shall represent the standard chemical potential of the rare gas with the same radius. If the ion A_i and the rare gas have the same electronic configuration, we can accept the following equality as being true:

$$\mu_i^{0(\mathrm{Gas})} = \mu_{GR}^{0(\mathrm{Gas})} \qquad [2.8]$$

Let $\mu_{GR}^{0(\mathrm{S})}$ denote the standard chemical potential (C = 1 mole/l) of the rare gas solvated. The standard chemical potential of the ion A_i solvated but not charged shall be equal to $\mu_{GR}^{0(\mathrm{S})}$. The chemical potential of the ion A_i solvated can then be written as:

$$\mu_i^{(\mathrm{S})} = \mu_{GR}^{0(\mathrm{S})} + \mathrm{R}T \ln \Gamma_i^{(\mathrm{S})} C_i \qquad [2.9]$$

The term $\mathrm{R}T \ln \Gamma_i^{(\mathrm{S})} C_i$ is due only to the electrical charge of the ion in the solvent S. Following the example set by Trémillon [TRÉ 71], we shall speak of the *activity coefficient of solvation*[3] of the ion A_i in the solvent S, denoted $\Gamma_i^{(S)}$.

The Gibbs molar energy of solvation will therefore be:

$$\Delta_{sol} G_{i(M)} = \mu_i^{(\mathrm{Liq})} - \mu_i^{0(\mathrm{Gas})} = \mu_{GR}^{0(\mathrm{Liq})} + \mathrm{R}T \ln \Gamma_i^{(\mathrm{S})} - \mu_{GR}^{0(\mathrm{Gas})} \qquad [2.10]$$

3 This value has been given different names by different authors. For Grünwald, it is the *degenerate activity coefficient*, and for Bates, the *medium effect coefficient*.

We can write this alternatively as:

$$\Delta_{sol} G_{i(M)}^{(S)} = \Delta_{sol} G_{GR}^{0(S)} + RT \ln \Gamma_i^{(S)} \qquad [2.11]$$

If we compare that relation to relation [2.6], we can assimilate the term relative to the rare gas with the non-electric term, and therefore write:

$$\Delta_{sol} G_{i(M)}^{(S)} - \Delta_{sol} G_{GR}^{0(S)} = \Delta_{élec} G_i^{0(S)} = RT \ln \Gamma_i^{(S)} \qquad [2.12]$$

Thus, by comparing with the energies involved in the cycle in Figure 2.3, we can write the following for one mole:

$$\Delta_{elec} G_i^{0(S)} = RT \ln \Gamma_i^{(S)} = N_a \left(W_d + W_c \right) \qquad [2.13]$$

We shall now calculate the terms W_d and W_c.

In fact, first of all, we shall calculate W_c, which is the energy required to charge a sphere whose radius is r_i, placed in a dielectric which has the constant D.

The energy stored in a dielectric S when we charge an ion is given by:

$$W_c = \frac{1}{2} \iiint_V \vec{D}.\vec{E}.\mathrm{d}v \qquad [2.14]$$

The integral is extended to the whole of the volume. The vectors $\vec{D}$ and $\vec{E}$ respectively represent the electric induction vector and the electric field vector.

If the dielectric can be considered to be perfect, we have:

$$\vec{D} = \varepsilon_0 D \vec{E}^2 \qquad [2.15]$$

Also, if ϕ denotes the electrical potential, we have:

$$\vec{E} = -\overrightarrow{grad \varphi} \qquad [2.16]$$

The charge energy is therefore written as:

$$W_c = -\frac{\varepsilon_0 D}{2} \iiint_V \vec{E}.\overrightarrow{grad\,\varphi}.\mathrm{d}v \qquad [2.17]$$

We know that:

$$div(\varphi\vec{E}) = -\overrightarrow{grad\,\varphi}.\vec{E} + \varphi.div\vec{E} \qquad [2.18]$$

By feeding this back into equation [2.17], we obtain:

$$W_c = \frac{\varepsilon_0 D}{2} \iiint_V \varphi.div\vec{E}.\mathrm{d}v - \frac{\varepsilon_0 D}{2} \iiint_V div(\varphi\vec{E}).\mathrm{d}v \qquad [2.19]$$

Poisson's equation outside of the ion immediately gives us:

$$div\vec{E} = 0 \qquad [2.20]$$

Let us apply the Green–Ostrogradsky theorem:

$$W_c = \frac{\varepsilon_0 D}{2} \iint_S \varphi.\overrightarrow{grad\,\varphi}.\vec{n}\,\mathrm{d}s \qquad [2.21]$$

The integration surface is the sphere of the ion whose radius is r_i. The potential ϕ generated by the charge ion ze in the homogenous medium with dielectric constant D is:

$$\varphi = \frac{ze}{4\pi\varepsilon_0 D r_i} \qquad [2.22]$$

Thus, by substituting back into relation [2.21], we find:

$$W_c = \left(\frac{ze}{4\pi\varepsilon_0 D}\right)^2 \iint_s \frac{1}{r}\,grad\frac{1}{r_i}.\vec{n}\,\mathrm{d}s \qquad [2.23]$$

In view of the sphericity of the ion, it makes sense to operate in spherical coordinates to calculate the integral. Thus, we have:

$$grad\frac{1}{r_i}=-\frac{\mathrm{d}r}{r_i^2} \tag{2.24}$$

The integral in relation [2.23] thus becomes, in spherical coordinates:

$$\int_0^{2\pi}\int_0^{\pi}\frac{\mathrm{d}r_i}{r_i}\sin\theta\,\mathrm{d}\theta\,\mathrm{d}\phi=\frac{4\pi}{r_i} \tag{2.25}$$

Thus, we have the charge energy of the ion:

$$W_c=\frac{(ze)^2}{8\pi\varepsilon_0 D r_i} \tag{2.26}$$

A similar calculation gives us the following for the discharge energy in a vacuum:

$$W_d=-\frac{(ze)^2}{8\pi\varepsilon_0 r_i} \tag{2.27}$$

By feeding back the results [2.26] and [2.27] into relation [2.13], we obtain:

$$\Delta_{elec}G_i^{0(S)}=\mathrm{R}T\ln\Gamma_i^{(S)}=\mathrm{N_a}\left(W_d-W\right)_c=-\frac{\mathrm{N_a}(ze)^2}{8\pi\varepsilon_0 r_i}\left(1-\frac{1}{D}\right) \tag{2.28}$$

As a dielectric constant varies between 2 and 100, we can see that this electrical contribution is always negative.

Using relations [2.6] and [2.28], the standard Gibbs molar energy of solvation becomes:

$$\Delta_{sol}G_i^{0(S)}=-\frac{\mathrm{N_a}(ze)^2}{8\pi\varepsilon_0 r_i}\left(1-\frac{1}{D}\right)-\Delta_{vdW}G_i^{0(S)} \tag{2.29}$$

We observe an affine variety of the Gibbs energy of solvation with the inverse of the radius of the ion and with the inverse of the dielectric constant.

Experience tells us that these two laws are not very rigorously respected. A number of reasons account for the differences between Born's model and reality.

The first approximation which we made is to assimilate the non-charged ion to a rare gas having the same radius. Whilst this approximation is not overly audacious in the case of ions of simple elements – i.e. normal cations and anions which do indeed have the same structure as the rare gas – but when they are charged, this is no longer true when we are dealing with cations of transition elements, and is all the more false with complex anions.

Then, we supposed the ion to be spherical, which still holds true for simple ions but differs notably from reality with complex ions containing multiple atoms.

We then supposed that the medium into which the ion is immersed is a homogeneous dielectric medium, characterized by a dielectric constant D which is identical at all points. However, evidently, due to the influence of the ion, in its vicinity, the polarization of the molecules of solvent causes an alteration of the dielectric constant, which only regains its normal value in the third corona around the ion (Figure 2.2).

Finally, we supposed, in employing Born's model, that the radius of the ion was independent of the solvent. We can see that the affine law in $1/r$ usually does not work. On the other hand, for simple spherical ions, we saw that the affine law in $1/r$ did work in water, as shown by Figure 2.4, on condition that we add 0.70µm to the radii of the cations and 0.30µm to the radii of the anions. Thus, the radius depends not only on the solvent, but also the first solvation layer produces an increase in the ion's apparent radius.

The non-electric term in relation [2.29] can be very significantly increased by the intervention of forces other than the van der Waals forces in the bond between the ion and the molecules of solvent. In particular, when there is the possibility for hydrogen bonds, that non-electric term may become very significant indeed, and its variations with ionic radius are not necessarily either negligible or affine in $1/r$. It is undeniable that in these cases, where the electrical part of the Gibbs energy of solvation is not necessarily predominant, it is illusory to try to use Born's model.

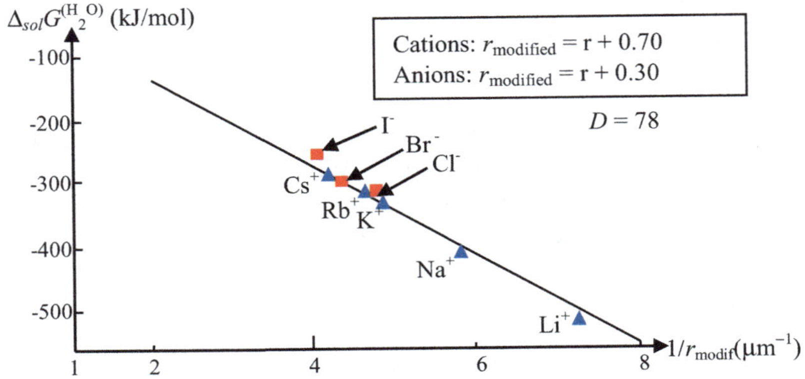

Figure 2.4. *Gibbs energy of solvation as a function of the modified ionic radii*

Finally, it is clear that for molecular solutes, Born's electrical contribution is zero.

2.4. Transfer of a solute from one solvent to another

For the study of solutions, we shall work with convention (III) – the molar solution.

We wish to compare the properties of a solute in a solvent S_1 with the same properties of the same solute in a different solvent S_2. In order to do so, it is helpful to compare the solute in the two solvents at the same chemical potential, i.e. the same energy level.

If, in the two solutions, the solute has the same chemical potential, this means that there is not the same amount of activity. Indeed, the reference chemical potentials, which are both defined at the same concentration of 1 mol/l, do not have the same value in the two solvents. We shall now introduce a new entity, called the *Gibbs energy of transfer* of the solute A_i from solvent S_1 to solvent S_2, defined by the relation:

$$\Delta_{S_1 \to S_2} G_i^0 = \mu_i^{0(S_2)} - \mu_i^{0(S_1)} \quad [2.30]$$

Because the chemical potential of the solute is the same in both solvents, we have:

$$\mu_i^{(S_2)} = \mu_i^{(S_1)} \qquad [2.31]$$

This equality, if we expand the expressions of the chemical potentials, is expressed by:

$$\mu_i^{0(S_2)} + 2.3RT\log\left|A_i\right|^{(S_2)} = \mu_i^{0(S_1)} + 2.3RT\log\left|A_i\right|^{(S_1)} \qquad [2.32]$$

From this relation, we deduce:

$$\log\frac{\left|A_i\right|^{(S_1)}}{\left|A_i\right|^{(S_2)}} = \frac{\mu_i^{0(S_2)} - \mu_i^{0(S_1)}}{2.3RT} = \frac{\Delta_{S_1\to S_2}G_i^0}{2.3RT} \qquad [2.33]$$

We shall define the *transfer activity coefficient* (TAC), written as ${}^{(S_1)}\Gamma_i^{(S_2)}$, by the relation:

$$\frac{\Delta_{S_1\to S_2}G_i^0}{2.3RT} = \log{}^{(S_1)}\Gamma_i^{(S_2)} \qquad [2.34]$$

By comparing relations [2.33] and [2.34], we can write:

$$\left|A_i\right|^{(S_1)} = {}^{(S_1)}\Gamma_i^{(S_2)}\left|A_i\right|^{(S_2)} \qquad [2.35]$$

That transfer activity coefficient will be greater than or less than 1 depending on whether the solute is more or less heavily solvated in solvent S_1 than in S_2.

Let us linger for a moment over relation [2.35]. We can say that it expresses an equilibrium of phase of the component A_i in the two solvents S_1 and S_2. This being the case, the transfer activity coefficient has the meaning of an equilibrium constant. We shall come back to this idea to measure the transfer activity coefficient between two immiscible solvents (see section 2.6.1.1). We can also express relation [2.35], saying that it establishes a scale of correspondence between the two solvents, thus enabling us to compare component A_i's behavior in all solvents. The activity values found in relation [2.35] correspond to the same chemical energy of component A_i. We can say that these two activity values are energetically equivalent. We establish a similar scale for the pH between the different solvents in Chapter 3 (see section 3.3.2).

In the case of molecular solvents at normal temperature, and of solutes which can also exist in aqueous solution, naturally, very often we express the properties of a solute in a solvent S in relation to its properties in water. By applying relation [2.35] to this case, we obtain:

$$\left|\mathrm{A}_i\right|^{(Water)} = {}^{(Water)}\Gamma_i^{(\mathrm{S})}\left|\mathrm{A}_i\right|^{(\mathrm{S})} \qquad [2.36]$$

The corresponding transfer activity coefficient, which – for the sake of simplicity – we shall henceforth denote by $\Gamma_i^{(\mathrm{Tr})}$, is such that:

$$\log {}^{(Water)}\Gamma_i^{(\mathrm{S})} = \log \Gamma_i^{(\mathrm{Tr})} = \frac{\mu_i^{0(\mathrm{S})} - \mu_i^{0(Water)}}{2.3\mathrm{R}T} \qquad [2.37]$$

Using relation [2.36], it is possible to express the activities of the solutes in any solvent, using a common scale where the reference state is that of the solute in water at the same reference concentration – i.e. 1 mol/l, or the infinitely-dilute solution.

In section 2.3.2, we discussed Born's model, in which the Gibbs energy of solvation was due only to electrical factors. Let us now evaluate the transfer activity coefficients in such a model. In other words, we can write that the transfer activity coefficient is the product of a van der Waals-type contribution and an electrical contribution, and it is the latter which we shall now evaluate.

We write the Gibbs energy of solvation (electric) of the same solute A_i in two solvents S_1 and S_2. According to relation [2.28], we have:

$$\Delta_{elec} G_i^{0(\mathrm{S}_1)} = \mathrm{R}T \ln \Gamma_i^{(\mathrm{S}_2)} \qquad [2.38]$$

$$\Delta_{elec} G_i^{0(\mathrm{S}_2)} = \mathrm{R}T \ln \Gamma_i^{(\mathrm{S}_2)} \qquad [2.39]$$

The Gibbs energy of transfer from solvent S_1 to solvent S_2 is therefore written:

$$\frac{\Delta_{S_1 \to S_2} G_i^0}{2.3RT} = \log \Gamma_i^{(S_2)} - \log \Gamma_i^{(S_1)} = \log \frac{\Gamma_i^{(S_2)}}{\Gamma_i^{(S_1)}} \quad [2.40]$$

By comparing this with relation [2.34], which is the definition of the transfer activity coefficient, we obtain:

$$^{(S_1)}\Gamma_i^{(S_2)} = \frac{\Gamma_i^{(S_2)}}{\Gamma_i^{(S_1)}} \quad [2.41]$$

By introducing relation [2.29], we are able to deduce:

$$2.3 \log {}^{(S_1)}\Gamma_i^{(S_2)} = \frac{N_a (ze)^2}{8\pi\varepsilon_0 r_i} \left(\frac{1}{D_1} - \frac{1}{D_2} \right) \quad [2.42]$$

Obviously, this is the electrical contribution to the total transfer activity coefficient.

This contribution, based on Born's model, obviously suffers from the same inaccuracies and the same disadvantages as those pointed out at the end of section 2.3.2.

In the case of comparison of a solvent S with water, relation [2.42] is written as:

$$2.3 \log \Gamma_i^{(Tr)} = \frac{N_a (ze)^2}{8\pi\varepsilon_0 r_i} \left(\frac{1}{D_S} - \frac{1}{D_{Eau}} \right) \quad [2.43]$$

According to this relation, when moving from water to a solvent with a lower dielectric constant (which is most often the case), the electrical effect must cause the transfer activity coefficient of the ion to increase.

2.5. Mean transfer activity coefficient of solvation of an electrolyte

For an electrolyte, it is obviously impossible to transfer one of the ions separately from one solvent into another. We can only calculate the Gibbs energy of total transfer. Therefore, we are led to define mean transfer activity coefficients.

Consider an electrolyte $A_{\nu_+}B_{\nu_-}$, whose ions carry the charges $z\nu_-$ and $z\nu_+$. We can distinguish between two contributions – one anionic and the other cationic. The total Gibbs energy of transfer will be:

$$\Delta_{Tr}G_t^0 = \nu_+\Delta_{Tr}G_A^0 + \nu_-\Delta_{Tr}G_B^0 \qquad [2.44]$$

We define the *mean transfer activity coefficient* by the following relation, inspired by relation [2.34]:

$$2.3\mathrm{R}T\log\Gamma_\pm = \frac{\nu_+\Delta_{Tr}G_A^0 + \nu_-\Delta_{Tr}G_B^0}{\nu_+ + \nu_-} \qquad [2.45]$$

Using relation [2.34] for the individual transfer activity coefficients, we find:

$$2.3\mathrm{R}T\log\Gamma_\pm = \frac{2.3\mathrm{R}T\nu_+\log\Gamma_A + 2.3\mathrm{R}T\nu_-\log\Gamma_B}{\nu_+ + \nu_-} \qquad [2.46]$$

From this, we deduce a relation that is very similar to relation [1.1], which defined the mean activity coefficient:

$$\Gamma_\pm = \left(\Gamma_+^{\nu_+}\Gamma_-^{\nu_-}\right)^{1/(\nu_+ + \nu_-)} \qquad [2.47]$$

For a 1-1 electrolyte, we have:

$$\Gamma_\pm = \left(\Gamma_+\Gamma_-\right)^{1/2} \qquad [2.48]$$

For the electrical contribution, using relations [2.46] and [2.43] we obtain:

$$2.3\mathrm{R}T\log\Gamma_{\pm}=\frac{1}{\nu_{+}+\nu_{-}}\frac{\mathrm{N_a}(e)^2}{8\pi\varepsilon_0 r_i}\left(\frac{1}{D_{\mathrm{S}}}-\frac{1}{D_{\mathrm{Water}}}\right)\left(\nu_{+}^2+\nu_{-}^2\right) \qquad [2.49]$$

which, in the case of a 1-1 electrolyte, gives us the expression:

$$2.3\mathrm{R}T\log\Gamma_{\pm}=\frac{\mathrm{N_a}(ze)^2}{16\pi\varepsilon_0 r_i}\left(\frac{1}{D_{\mathrm{S}}}-\frac{1}{D_{\mathrm{Water}}}\right)\left(\frac{1}{r_{\mathrm{A}}}+\frac{1}{r_{\mathrm{A}}}\right) \qquad [2.50]$$

Of course, this last relation contains the same faults and the same inaccuracies as its counterparts encountered above.

2.6. Experimentally determining the transfer activity coefficient of solvation

We wish to determine the transfer activity coefficient of a solute A_i between two solvents S_1 and S_2.

In determining the transfer activity coefficients, we must distinguish between two cases: that of molecular solutes, which does not pose a major problem, and that of ionic solutes. In the latter case, we either determine the mean transfer activity coefficient or we attempt to find an individual transfer activity coefficient for the ions.

In the methods employed, we distinguish thermochemical methods from electrochemical ones, which involve an electrochemical cell. The latter category of methods is discussed in detail in Chapter **7**, which is devoted to the thermodynamics of such cells.

With all thermochemical methods, we need to know the relation between the activity coefficient of component A_i in question and its concentration in each of the solvents. Either that relation is determined in advance using conventional methods for determining chemical activities, or the solutions are sufficiently dilute to allow us to treat the activity as being equal to the concentration.

The problem now is how to determine the concentrations of component A_i in each of the solvents. These concentrations are such that they produce relation [2.35] (or [2.36], if one of the solvents is water) between the activity values.

2.6.1. *Determining the activity coefficient of a molecular solute*

Thermochemical methods are based on the comparison of two solutions, either directly or with a third phase containing the same component, with known activity and at equilibrium, separately, with the two solutions being studied. Usually, this third phase is the pure substance A_i. Then, the equilibria in question are equilibria of solubility. Indeed, if each of the solutions is placed at equilibrium with the pure substance A_i, this means that the chemical potential of A_i in those solutions is the same as that of A_i in the pure substance. Thus, relation [2.31] will be respected between the two solutions.

2.6.1.1. *Use of equilibrium of sharing between immiscible solvents*

If the two solvents are not miscible with one another, we merely need to introduce a small amount of component A_i into one of the solvents, mix the solvents and then, when equilibrium is achieved, dose A_i into each solvent. We can use small quantities of A_i so as to be able to treat the concentration and the activity as being one and the same thing. It is easy to calculate the transfer activity coefficient between the two solvents, using the following relation, derived from relation [2.36], in the case that one of the solvents is water:

$$\Gamma_i^{(\mathrm{Tr})} = \frac{C_{equil}^{(\mathrm{Water})}}{C_{equil}^{(\mathrm{S}_2)}} \qquad [2.51]$$

If the water is replaced by a different solvent, we use relation [2.35].

2.6.1.2. *Use of the comparison of the solubilities in the two solvents*

In this method, we dissolve the solute, to the point of saturation, in each of the two solvents separately, at the same temperature. Given that, in both cases, the solution is at equilibrium with the pure substance, the chemical potential of each solute is equal to the chemical potential of the same component in the pure state, and thus they are equal to one another. Hence, the chemical potentials of the solute in the two solutions obey relation [2.31]. Therefore, we simply need to measure the corresponding activities in each solution and apply relation [2.35] or [2.36] as required.

NOTE.– If the solubility values are low, we can treat the activities and concentrations as being one and the same.

Obviously, we must take the precaution of verifying that the pure compound at equilibrium with the solution is indeed the same in both cases.

If the solute is a solid or a liquid in the pure state, la method presents no difficulty at all. If this compound is in gaseous form in the pure state, the equilibria of saturation need to be achieved at the same partial pressure of the gas in question.

2.6.2. *Determination of the mean transfer activity coefficient of a strong electrolyte*

In the case of strong electrolytes which dissociate in solution, without considering the parasitic reactions of solvolysis or complexation, we can use two methods which will both yield the mean transfer activity coefficient as defined in section 2.5.

The first method will be the same as before: we use the notion of solubility of the electrolyte in each of the two solvents (see section 2.6.1.2). In order to determine the mean coefficient, we first determine the mean activity in the two solutions if the solubilities are not sufficiently low to enable us to treat the activity and the concentration as being equal.

The second method uses the equilibrium of an electrochemical cell, and will be described in section 7.7.12.

2.6.3. *Evaluation of the individual transfer activity coefficient of an ion.*

It is impossible to imagine experimental determination of an ion's intrinsic transfer activation coefficient, because it is not possible to isolate that ion.

The idea is to evaluate the coefficient of activation of intrinsic transfer of an ion by an *extra-thermodynamic hypothesis*.

We proceed on the basis of knowing either the total standard Gibbs energy of transfer, or the mean coefficient of transfer activation of the electrolyte. These values have been measured either by an electrochemical method or by a thermochemical method. We shall now examine the three hypotheses reported in the literature, which appear to be the most reliable.

2.6.3.1. *Pleskow's hypothesis*

Although it is probably not the most reliable of methods, this one has the advantage of being the longest-standing (established in 1947).

The approach is to accept that there is an ion for which the activity coefficient of solvation is equal to 1 for all solvents (standard Gibbs energy of transfer of 0). If such an ion exists, it must probably have a high radius and only a small charge ($z = 1$). When we look at the periodic table, the cesium ion (Cs^+) seems an obvious choice for this experiment. In Pleskow's work, as cesium salts are rare, he chose to use the rubidium ion (Rb^+) instead. Since then, tests have been carried out with cesium salts, and the results are comparable to those obtained with the rubidium salts.

In order to improve the accuracy of Pleskow's hypothesis, instead of the unit value for $\Gamma_i^{(\mathrm{Tr})}$ of cesium, we can use the value given by Born's model (see relation [2.43]) for cesium chloride. Attempts to do this have shown that the correction was not hugely significant.

Thus, we determine the value of the mean activity coefficient, between water and the chosen solvent, for a salt such as chloride. In order to do so, we use a thermochemical method (see section 2.6.2) or an electrochemical method (see section 7.7.13.1). We then directly obtain the following, thanks to relation [2.48]:

$$\Gamma_{\pm} = \left(\Gamma_{-}\right)^{1/2} \qquad [2.52]$$

Once we have thus found the transfer activity coefficient of a positive ion for the solvent used, it is easy, little by little, to determine that of any given ion. To do so, we measure $\Gamma_{\pm}$ of a salt containing an ion whose transfer activity coefficient has already been determined, and apply relation [2.47].

2.6.3.2. *Strehlow's hypothesis*

Strehlow uses a potentiometric method (see section 7.7.13.2) and determines the total Gibbs energy of transfer of a set formed of a cation M^+ and a redox pair formed of ferrocene and a ferricinium cation. Ferrocene, which has the chemical formula $(C_6H_5)_2Fe$, is a sandwich structure formed of a cation Fe^{2+} between two cyclopentadienyl anions. The ferricinium ion has the same structure, but the ferric ion is in the form Fe^{3+}.

The total Gibbs energy of transfer has the form:

$$\Delta_{\text{Tr}} G_{total} = \Delta_{Tr} G_{\text{M}^+} + \left(\Delta_{Tr} G_{\text{ferrocene}} - \Delta_{Tr} G_{\text{ferricinium}}\right) \qquad [2.53]$$

In the author's view, the two terms $\Delta_{\text{Tr}} G_{\text{ferrocene}}$ and $\Delta_{\text{Tr}} G_{\text{ferricinium}}$ are very similar to one another, the only difference lying only in the ionic nature of the latter term. As the electrostatic corrections which it is possible to make are negligible, we deduce that measuring $\Delta_{\text{Tr}} G_{total}$ directly gives us $\Delta_{\text{Tr}} G_{\text{M}^+}$, which is why we calculate the transfer activity coefficient of the cation M^+. Once this value is obtained for one cation, it can be determined conventionally for other ions, little by little.

The ferrocene–ferricinium pair can also be replaced by the cobaltocene–cobalticinium pair, which has the same structure, with cobalt replacing the iron.

The results obtained are very similar to those given by the rubidium method.

2.6.3.3. *Grunwald and Popovych's hypothesis*

Grunwald and Popovych postulate that for certain salts, formed of an anion and a very similar cation, the activity coefficients of transfer between two solvents would always be equal. The authors propose to use the couple formed of the tetraphenylarsonium cation φ_4As^+ (or tetraphenylphosphonium φ_4P^+) and the tetraphenylborate anion φ_4B^-. By measuring the solubility, we deduce the transfer activity coefficient $\Gamma_\pm$, giving us the individual coefficients calculated by:

$$\Gamma_\pm = \Gamma_+^{1/2} = \Gamma_-^{1/2} \qquad [2.54]$$

For example, for the transfer between acetonitrile and water, we find $\Gamma_+ = \Gamma_- = -6.2$.

Armed with this value, we then determine the solubilities of tetraphenylarsonium iodide and silver tetraphenylborate, which respectively give us the coefficients of the two I^- and Ag^+ ions. Then, little by little, we find the values of all of the ions.

The results obtained by this method are very similar to those obtained by Strehlow's method, as shown by Table 2.2, which compares the results of these two methods for the transfer of halide ions between water and acetonitrile. This matching of the results lends a certain credibility to the two methods.

Evaluation method	Cl^-	Br^-	I^-
Grunwald–Popovych	7.9	5.4	3.0
Strehlow	5.9	6.0	4.1

Table 2.2. *Comparison between the Grunwald–Popovych method and Strehlow's method for evaluating the transfer activity coefficients of halide ions in acetonitrile in relation to water [TRÉ 71]*

Whilst the use of multiple methods lends a certain degree of veracity to the values obtained, values found by the use of only one method exhibit the advantage of being consistent. Table 2.3 shows the values obtained for different ions, all measured by Grunwald and Popovych's method and the use of the supposed property of the tetraphenylarsonium and tetraphenylborate ions (see section 2.6.3.3)

Ions		Solvents		
		Acetonitrile	Ethanol	DMF
Cations	Ag^+	-3.4	0.8	-4.6
	K^+	-0.4	0.0	-2.5
	Ca^{2+}	-0.5	-0.1	-2.4
Anions	Cl^-	7.9	1.5	8.3
	Br^-	5.4	1.1	6.3
	I^-	3.0	0.5	3.4
	$CH_3CO_2^-$	10.3	1.9	11.4

Table 2.3. *Values of the transfer activation coefficients in relation to water for a number of ions, measured using Grunwald and Popovych's method [TRÉ 71]*

Yet we must not lose sight of the fact that all these values are estimated rather than measured, and are based on extra-thermodynamic hypotheses which, though they are certainly plausible, are not proven.

2.7. Relation between the constants of the same equilibrium achieved in two different solvents

It is evident that when the solvent takes part in a chemical reaction, its standard chemical potential enters into the expression of the equilibrium constant, and therefore we can say that the solvent plays a chemical role in determining the equilibrium constant.

We shall now examine a more physical influence of the solvent on an equilibrium constant – an influence which is due to the phenomenon of solvation, independently of the chemical role discussed above.

2.7.1. *General relation of solvent change on an equilibrium constant*

Let us consider an equilibrium very generally written with the usual convention in terms of the signs of the stoichiometric coefficients ν_I, which are positive for the right-hand side of the equilibrium and negative on the left-hand side.

$$0 = \sum_i \nu_i A_i \qquad [2R.5]$$

The equilibrium constant in a solvent S_j is defined on the basis of the standard Gibbs energy of the reaction in the same solvent, as follows (see la relation [A.2]):

$$K^{(S_j)} = \exp - \frac{\Delta_r G^0_{S_j}}{RT} \qquad [2.55]$$

The difference between the two logarithms of that constant in two solvents S_1 and S_2 will therefore be:

$$\Delta \ln {}^{(S_2)}K^{(S_1)} = \ln K^{(S_2)} - \ln K^{(S_1)} = \frac{\Delta_r G^0_{(S_1)}}{RT} - \frac{\Delta_r G^0_{(S_2)}}{RT} \qquad [2.56]$$

The difference in the right-hand side of the two standard Gibbs energies in the two solvents reveals the Gibbs energy of transfer from solvent S_2 to solvent S_1. That Gibbs energy is expressed as a function of the transfer activity coefficients, and by application of relation [2.34], we obtain:

$$\Delta \ln {}^{(S_2)}K^{(S_1)} = \Delta\left(\Delta_{Tr}\ {}^{(S_2)}G_i^{0\ (S_1)}\right) = -2.3\sum_i \nu_i\ {}^{(S_2)}\log \Gamma_i^{(S_1)} \qquad [2.57]$$

For example, for the equilibrium constant in a solvent S in relation to water, we obtain:

$$\Delta \ln {}^{(S)}K^{(\text{Water})} = -2{,}3\sum_i \nu_i \log \Gamma_i^{(\text{Tr})} \qquad [2.58]$$

Let us choose an example of the application of relation [2.58] for a decomposition reaction, written as:

$$\text{AB}_n = \text{A} + n\text{B} \qquad [2R.6]$$

Relation [2.58] directly yields:

$$\ln\ K^{(S)} - \ln\ K^{(\text{Water})} = \Gamma_{\text{AB}_n}^{(\text{Tr})} - \Gamma_{\text{A}}^{(\text{Tr})} - n\Gamma_{\text{B}}^{\text{Tr}} \qquad [2.59]$$

Thus, we can see that, if the increase in solvation of compounds A and B is greater than the increase in solvation of AB_n, the equilibrium constant in the solvent S is greater than its value in water.

2.7.2. *Influence of the dielectric constant of the solvent on the equilibrium constant of an ionic reaction*

When ions are involved in the equilibrium, a certain portion of the variations of the constant with the solvent can be attributed to the electrostatic fraction of the transfer activity coefficients. That electrostatic part is, as we know, insufficient to correctly calculate these coefficients. However, that part does enable us to account for the influence of the dielectric constant of the solvents. In order to do so, we rely on relation

[2.42] and its particular application [2.43] when one of the two solvents is water.

To study this influence of the dielectric constant experimentally, we can use as solvents mixtures of varying compositions of two miscible solvents whose dielectric constants are substantially different. Thus, we can easily use mixtures of water and ethyl alcohol, whose respective dielectric constants are 78 and 25 (Table 2.1), or mixtures of water and dioxane ($D = 9.5$).

We shall now look at two types of reaction in order to demonstrate the thinking process and the nature of the results obtained.

2.7.2.1. *Case of isoionic equilibria*

An equilibrium is said to be *isoionic* if it does not alter the number and charge of the ions. Such an equilibrium is one where the number of ions is the same on both sides.

For instance, the equilibria [2R.7] and [2R.8] shown below are isoionic equilibria.

$$AB^{+} = A^{+} + B \qquad [2R.7]$$

$$AB^{+} + CD^{2+} = AC^{+} + BD^{2+} \qquad [2R.8]$$

Indeed, in the first example, we have one ion on the left-hand side and one ion on the right, both of which obviously carry the same charge. In the second example, we have two ions on the left and two on the right of the equilibrium, with the charges on the ions being the same – namely one ion with the charge +1 and one with the charge +2 on both sides of the equilibrium.

We shall now examine the influence of the dielectric constants of the solvents on the variation of the equilibrium constant when moved from water to a solvent S to perform reaction [2.7].

The electrostatic contribution to the variation of the equilibrium constant can easily be calculated using relations [2.58] and [2.43].

Indeed, by using relation [2.58], as only the ions make an electrostatic contribution, we can write:

$$\Delta_{\text{elect}} \ln {}^{(\text{S})}K^{(\text{Water})} = 2.3\left(\log \Gamma^{(\text{Tr})}_{\text{AB}^+} - \log \Gamma^{(\text{Tr})}_{\text{A}^+}\right) \qquad [2.60]$$

By expressing the coefficients $\Gamma^{(\text{Tr})}_{\text{AB}^+}$ and $\Gamma^{(\text{Tr})}_{\text{A}^+}$ by relation [2.43], we find:

$$\Delta_{\text{elect}} \ln {}^{(\text{S})}K^{(\text{Water})} = \frac{\text{N}_{\text{a}}(e)^2}{8\pi\varepsilon_0}\left(\frac{1}{D_{\text{S}}} - \frac{1}{D_{\text{Water}}}\right)\left(\frac{1}{r_{\text{AB}^+}} - \frac{1}{r_{\text{A}^+}}\right) \qquad [2.61]$$

We can see that the lowering of the dielectric constant is favorable to the larger ion.

If the two ions have very similar ionic radii, the dielectric constant has no impact on the equilibrium constant. Indeed, this is what we see for the dissociation constant of AH^+-type cationic acids, such as NH_4^+.

2.7.2.2. *Equilibrium with increase in the number of ions*

Evidently, if the equilibrium is no longer isoionic, the conclusions about the influence of the dielectric constant on the equilibrium constant are more complex, and the expression needs to be found in each individual case.

By way of example, we shall now look briefly at the influence of the dielectric constant on the equilibrium constant between water and a solvent when the reaction takes place with an increase in the number ions, as in the example of reaction [2R.9], which we write as:

$$\text{AB} = \text{A}^+ + \text{B}^- \qquad [2\text{R}.9]$$

Similarly as in section 2.7.2.1, we apply relations [2.58] and [2.43] in turn. This calculation gives us:

$$\Delta_{\text{elect}} \ln {}^{(\text{S})}K^{(\text{Water})} = \frac{\text{N}_{\text{a}}(e)^2}{8\pi\varepsilon_0}\left(\frac{1}{D_{\text{S}}} - \frac{1}{D_{\text{Water}}}\right)\left(\frac{1}{r_{\text{B}^-}} + \frac{1}{r_{\text{A}^+}}\right) \qquad [2.62]$$

A relation such as this could be envisaged, for instance, in the case of certain molecular or anionic acids.

Thus, we can see that, when we move from a given solvent to another whose dielectric constant is lower, then by comparing relations [2.61] and [2.62], we can predict a drop in the acidity constant of molecular and anionic acids in comparison to cationic acids.

It must be remembered that we have only evaluated the electrostatic part of the variation of the dielectric constant. This part is sometimes so slight that, in experiments, we see an overall effect that is contrary to the electrical effect between two solvents.

3

Acid/Base Equilibria

Reactions involving ions generally take place in solution. The ions are then relatively dilute, so we can usually consider the activity of the solvent (often water) to be equal to 1. Ions also come into play in other media, such as non-aqueous solvents, but also in molten-salt environments.

Certain reactions involving ions also take place in a solid medium. We saw an example of such reactions in volume 3 of this collection [SOU 15c].

Ionic equilibria can be classified into two categories:

– *homogeneous equilibria* including acid–base equilibria, which we are looking at in this chapter, equilibria of formation and dissociation of complexes and redox equilibria, which will be discussed in Chapter 4;

– *heterogeneous equilibria*, which are essentially linked to the phenomena of solubilities accompanied by dissociations, which we shall examine in Chapter 5.

This chapter and the next essentially involve species diluted in a solvent. In defining the activities, we choose to take the 1 mole/l solution as a reference solution. We know that in this case, the chemical potential of the reference state is practically equal to that defined in reference (II) – the infinitely-dilute solution.

$$\mu_i^{0(III)} = \mu^{0(\infty)} \tag{3.1}$$

The reactions in solution that we are examining here are all exchange reactions. Thus, they are characterized by zero variation of the number of moles $\Delta\nu_t$. From this, if we look at relation [A.11], it follows that the equilibrium constant is the same whether we are dealing with the molar fractions or the concentrations. Thus, we shall have:

$$K^{(\mathrm{II})} = K^{(\mathrm{IJI})} \qquad [3.2]$$

Hence if in the course of certain applications in highly-dilute solutions we replace the activities with the concentrations, the equilibrium constants would not change.

3.1. Definition of acids and bases and acid–base reactions

We shall now look at acids and bases in solution. For the time being, we shall only consider the use of molecular protic solvents, in which Brønsted's acidity theory is perfectly applicable, and for which Lewis' theory is of no use. It is understood that the protons are attached to pairs of free electrons, making de Brønsted's approach a particular case of Lewis' approach (see section 3.13).

Remember that an acid is a molecule capable of ceding protons to another molecule and that a base is a molecule capable of capturing protons from another molecule. Thus, for each acid, there is a corresponding base, and an acid–base pair is represented by the reaction:

$$\text{Acid} = \text{Base} + \text{H}^+ \qquad [3\text{R}.1]$$

It is understood that the above reaction [3R.1] is meaningless on its own, because the lone proton is never stable. Thus, what makes chemical sense is reactions between the acid from one acid–base pair and the base from a different pair. Such a reaction constitutes an acid–base reaction, which we write as:

$$\text{Acid 1} + \text{Base 2} = \text{Acid 2} + \text{Base 1} \qquad [3\text{R}.2]$$

An equilibrium such as this is characterized by an equilibrium constant which is written in terms of the activities of the components:

$$K_2 = \frac{|\text{Acid 2}|\,|\text{Base 1}|}{|\text{Acid 1}|\,|\text{Base 2}|} \qquad [3.3]$$

Because we are dealing with an exchange reaction, if the solution is very dilute, we can replace the activities with the concentrations, and write:

$$K_2 = \frac{[\text{Acid 2}]\,[\text{Base 1}]}{[\text{Acid 1}]\,[\text{Base 2}]} \qquad [3.4]$$

We know that this equilibrium constant may depend on the solvent.

3.2. Ion product of an amphiprotic solvent

Amphiprotic solvents act both as proton donors (i.e. acids) and as proton acceptors (i.e. bases). These solvents give rise to an acid-base equilibrium between a molecule of solvent acting as an acid and a second molecule of solvent acting as a base. For a solvent SH, that equilibrium, called the *autoprotolysis equilibrium*, is written as:

$$2\text{HS} = \text{H}_2\text{S}^+ + \text{S}^- \qquad [3R.3]$$

Generally, the progress rate at equilibrium is very small, so for that equilibrium, the law of mass action is written:

$$K_{\text{SH}} = |\text{H}_2\text{S}^+|\,|\text{S}^-| \qquad [3.5]$$

This constant is called the *ion product of the solvent*. Its value differs greatly from one solvent to another. Table 3.1 shows the values equilibrium of autoprotolysis of a number of solvents, and their ion products.

Solvent	*Equilibrium of autoprotolysis*	*Ion product*
Acetic acid	$2CH_3COOH = CH_3COH_2^+ + CH_3COO^-$	10^{-13}
Water	$2H_2O = H_3O^+ + OH^-$	10^{-14}
Ethanol	$2C_2H_5OH = C_2H_5OH^{2+} + C_2H_5O^-$	10^{-19}
Ammonia	$2NH_3 = NH_4^+ + NH_2^-$	10^{-32}

Table 3.1. *Equilibria of autoprotolysis and ion products of a few solvents*

The neutral point of a solvent is determined by half its pK_i.

NOTE.– The autoprotolysis reaction is also an exchange reaction.

Obviously, ammonia, which is practically not a donor, has a very low ion product. In Chapter 7, we shall see a method for determining the ion product of a solvent (see section 7.7.6).

3.3. Relative strengths of acids and bases

Acids are classified by a value called the *acidity constant*, which depends on the solvent in which the acid–base reactions take place.

3.3.1. *Definition of the acidity constant of an acid*

When an acid AH is dissolved in a solvent SH, an initial reaction occurs, which is called *ionization* or *protolysis*:

$$\mathrm{AH} + \mathrm{HS} = \left\{\mathrm{A}^-, \mathrm{H_2S}^+\right\} \qquad [3R.4]$$

The corresponding equilibrium constant is written:

$$K_i = \frac{\left|\left\{\mathrm{A}^-, \mathrm{H_2S}^+\right\}\right|}{\left|\mathrm{A}^-\right|} \qquad [3.6]$$

Thermodynamically, when studying equilibria, it is always possible to break that ionization down into two different stages, where the respective equilibria would be:

$$\mathrm{AH} = \mathrm{A}^- + \mathrm{H}^+ \qquad [3R.4a]$$

and

$$\mathrm{A}^- + \mathrm{H}^+ + \mathrm{HS} = \left\{\mathrm{A}^-, \mathrm{H_2S}^+\right\} \qquad [3R.4b]$$

The corresponding laws of mass action would be:

$$K_{AH} = \frac{|A^-|\,|H^+|}{|AH|} \qquad [3.7]$$

$$K_p = \frac{|\{A^-,H_2S^+\}|}{|A^-|\,|H^+|} \qquad [3.8]$$

The equilibrium constant for the reaction [3R.4] is obviously linked to those for the equilibria [3R.4a] and [3R.4b]. Indeed, we have:

$$K_i = K_{AH}K_p \qquad [3.9]$$

The equilibrium [3R.4a] is intrinsic to the acid AH, and independent of the solvent. It quantifies the acid's ability to release protons. The higher the value of K_{AH}, the further to the right the equilibrium [3R.4a] shifts. We say that K_{AH} quantifies the intrinsic strength of the acid AH.

The sum of equilibria [3R.4a] and [3R.4b] (that is, equilibrium [3R.4]) indicates the ionizing power of the solvent SH. The more ions $\{A^-,H_2S^+\}$ are created, the stronger the effect of the solvent. This is also a mark of how strongly the solvent captures the proton. The higher the value of the constant K_i, the more basic the solvent is.

After ionization comes the dissociation reaction, which is written as:

$$\{A^-,H_2S^+\} = A^- + H_2S^+ \qquad [3R.5]$$

The corresponding dissociation constant is written:

$$K_D = \frac{|A^-|\,|H_2S^+|}{|\{A^-,H_2S^+\}|} \qquad [3.10]$$

The larger the value of K_D, the more ion pairs are dissociated.

We define the *acidity constant* K_a of the acid AH by the relation:

$$K_a = \frac{|A^-||H_2S^+|}{|\{A^-,H_2S^+\}|}\frac{|\{A^-,H_2S^+\}|}{|AH|+|\{A^-,H_2S^+\}|} \qquad [3.11]$$

The first fraction on the right-hand side of relation [3.11] is none other than the dissociation constant; the second represents the fraction of ions A^- solubilized. It is a measure of how basic the solvent is. In view of the equilibrium [3R.4], the solubilized fraction α_i is written:

$$\alpha_i = \frac{K_i}{1+K_i} \qquad [3.12]$$

The higher the value of α_i, the stronger K_i is, and therefore the more basic the solvent is, given that K_i measures the acidity of the solvent for the proton.

By feeding expression [3.10] back into the definition of the acidity constant [3.11], we obtain:

$$K_a = K_D\alpha_i \qquad [3.13]$$

From this, we derive:

$$pK_a = -\log K_a = pK_d - \log \alpha_i \qquad [3.14]$$

We can deduce from this that pK_a decreases as the acid dissociates. This result is well known in the case of aqueous solutions.

NOTE.– In light of relations [3.7], [3.8], [3.9] and [3.12], we can write:

$$K_a = \frac{K_iK_D}{1+K_i} = \frac{K_D}{1+\dfrac{1}{K_i}} \qquad [3.15]$$

According to the definition [3.11], the value K_a is not an equilibrium constant. However, as shown by relation [3.15], it is defined on the basis of such constants, so it is a thermodynamic value. The acidity constant K_a

depends on the temperature and of course on the acid AH, but also on the nature of the solvent SH.

For a totally-dissociated acid, we find:

$$\mathrm{p}K_a \cong \mathrm{p}K_D \qquad [3.16]$$

In highly-dissociative solvents (dielectric constant $D > 40$), with equal charges, the concentration of ion pairs $\left\{\mathrm{A}^-,\mathrm{H_2S}^+\right\}$ is very low. Thus, the activity of the ion pair becomes negligible in comparison to that of the non-dissociated acid AH, and we obtain the relation:

$$K_a = \frac{\left|\mathrm{A}^-\right|\left|\mathrm{H_2S}^+\right|}{\left|\mathrm{AH}\right|} \qquad [3.17]$$

This relation classically applies in the case of water, which has a dielectric constant of around 80. Relation [3.17] is absolutely rigorous in the case of a molecular solvent (such as water) and an acid whose base is electrically neutral (such as NH_3), because then we are no longer dealing with an ion pair.

In water, the pK_a of fully-dissociated acids is negative, but this is not the case in all solvents. In solvents which do not have high dissociative power, pK_a is no longer close to pK_D, which is negative (because of the term in K_i). For example, perchloric acid is totally ionized in acetic acid, but the pK_a of the perchlorate/perchloric acid couple is near to 5.

3.3.2. *Protic activity in a solvent*

By examining relations [3.8] and [3.10], we note that we can write:

$$K_{\mathrm{H_2S}^+} = \frac{\left|\mathrm{H}^+\right|}{\left|\mathrm{H_2S}^+\right|} = \frac{1}{K_p K_D} \qquad [3.18]$$

$K_{\mathrm{H_2S}^+}$ is, in fact, the constant of the equilibrium:

$$\mathrm{H_2S}^+ = \mathrm{H}^+ + \mathrm{HS} \qquad [3R.\ 6]$$

This is the intrinsic constant of the acid H_2S^+ (application of relation [3.7] to the acid H_2S^+). As with all constants of this type, we cannot directly find this equilibrium, or therefore its constant. However, if we apply this expression [3.18] where water is used as the solvent, we would write:

$$K_{H_3O^+} = \frac{|H^+|}{|H_3O^+|} = 1 \qquad [3.19]$$

Given that we do not have access to that constant, because the proton does not exist in appreciable quantity in water, the reversibility of the different equilibria suggests that the activity of the proton is proportional to that of H_3O^+. By convention, we posit that this coefficient of proportionality is equal to 1, so $pK_{H_3O^+} = 0$.

We speak of the *protic activity* of an aqueous solution for the activity of H_3O^+ in that solution. We can see that the pH is simply the opposite of the logarithm of the protic activity.

In the protic solvent SH, using relation [2.36], we define the protic activity a_{SH}^+ of the solution by the relation:

$$a_{SH}^+ = \Gamma_H^{(Tr)} |SH_2^+| \qquad [3.20]$$

$\Gamma_H^{(Tr)}$ is the transfer activity coefficient between water and the solvent SH.

Hence, we can compare the protic activities of an aqueous solution and a solution in the solvent SH, and the protic activities of two solutions in any two given solvents by expressing all of them in relation to water.

We let $pH^{*(SH)}$ denote the value defined by:

$$pH^{*(SH)} = -\log a_{SH}^+ \qquad [3.21]$$

That is the pH which, in water, the solution in SH with the same protic activity would have.

Using relation [3.20] in logarithmic form, we find:

$$pH^{*(SH)} = -\log \Gamma_{H}^{(Tr)} - \log\left|SH_2^+\right| = p\left(SH_2^+\right) - \log \Gamma_{H}^{(Tr)} \qquad [3.22]$$

Using this relation, we are able to establish an equivalent pH scale for a solvent other than water, and thus establish a common pH scale between the two solvents.

NOTE.– Certain authors speak of the value $p\left(SH_2^+\right)$ as the pH of the solution in the solvent SH. We shall not use this connotation here, so as not to confuse that pH with the protic activity of the solution which, on its own, can be used to compare behavior in two different solvents.

3.4. Direction of acid–base reactions, and domain of predominance

Consider an acid AH which reacts with a base B according to the reaction:

$$AH + B = A^- + BH^+ \qquad [3R.7]$$

Depending on the value of the corresponding equilibrium constant K_{AB}, the progress rate at equilibrium will be greater or lesser. We can define two domains:

– if the constant K_{AB} is greater than 1, the progress rate at equilibrium will be higher than 0.5, and the medium will contain mainly the base A^- and the acid BH^+. We say that this domain is the domain of predominance of A^- and BH^+;

– if the constant K_{AB} is less than 1, the progress rate at equilibrium will be less than 0.5, and the base B and acid AH will be in the majority in the medium. We say that this domain is the domain of predominance of AH and B.

However, in a sufficiently ionizing solvent, such that relation [3.11] is applicable, the constant K_{AB} is the ratio between the acidity constants of acids AH, K_{aA} and BH^+, K_{aB}, which is:

$$K_{AB} = \frac{K_{aA}}{K_{aB}} \qquad [3.23]$$

On a pH axis, the acidity constants are the constants expressed in relation to the solvent water.

Using relations [3.17] and [3.20], we obtain:

$$K_{aA}^{(SH)} = \frac{|A^-||SH_2^+|}{|AH|} = \frac{|A^-|\dfrac{|A^-|}{|AH|\Gamma_H^{(Tr)}}}{|AH|} = \frac{K_{aA}^{*(SH)}}{\Gamma_H^{(Tr)}} \qquad [3.24]$$

$K_{aA}^{*(SH)}$ is the acidity constant in the solvent SH, expressed on a scale where the origin is given by the pK_a of the equilibrium H_3O^+/H_2O. This is the acidity constant in the solvent SH, expressed on the scale of the pK values of water.

From relation [3.24], we deduce:

$$K_{aA}^{*(SH)} = K_{aA}^{(HS)} \Gamma_H^{(Tr)} \qquad [3.25]$$

That relation tells us that the $pK_{aA}^{(SH)}$ and $pK_{aA}^{*(SH)}$ scales can be deduced from one another by a simple translation of vector $\Gamma_H^{(Tr)}$.

It is important not to confuse the pK_a of an acid in water ($pK_{aA}^{(H_2O)}$) with the pK_a of the same acid in a different solvent SH, expressed on the scale in relation to water: $pK_{aA}^{*(SH)}$. Therefore, we write:

$$pK_{aA}^{(H_2O)} \neq pK_{aA}^{*(SH)}$$

This difference is clearly illustrated in Figure 3.7.

Let us now consider the case where the constant K_{AB} is greater than 1, meaning that $pK_{aA}^{(SH)} < pK_{aB}^{(SH)}$ (Figure 3.1). On the pH axis, we can write the two values $pK_{aA}^{*(SH)}$ and $pK_{aB}^{*(SH)}$, calculated by way of relations similar to equation [3.25]. These two values are placed in the same order and at the same distance apart as the values $pK_{aA}^{(SH)}$ and $pK_{aB}^{(SH)}$.

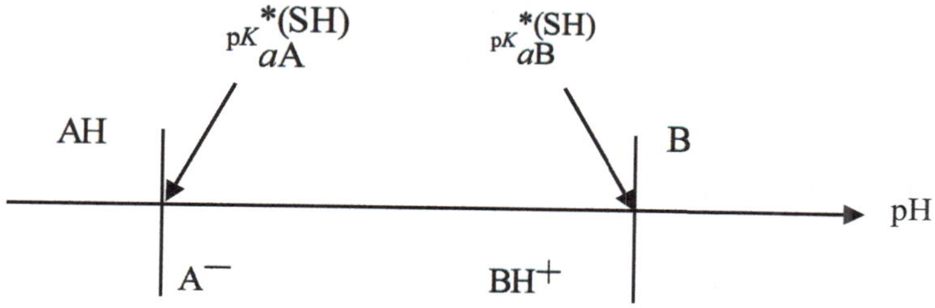

Figure 3.1. *Domain of predominance of the compounds in an acid–base reaction*

To the left of the point at abscissa $pK_{aA}^{*(SH)}$, the compound AH is predominant. To the right of that point, it is the compound A^- which is predominant.

To the left of the point at abscissa $pK_{aB}^{*(SH)}$, the compound BH^+ is predominant. To the right of that point, it is the compound B which is predominant.

Note that the acid AH and the base B do not coexist (which is unsurprising, given that $K_{AB} > 1$); we say that AH and B are *incompatible*. On the other hand, the base A^- and the acid BH^+ are compatible, because they coexist in the same domain between the two points.

NOTE.– On the diagram, we can see that to the left of the abscissa $pK_{aA}^{*(SH)}$, the two acids AH and BH^+ are compatible, which is obvious because no reaction occurs when we mix two acids.

We shall see in section 3.7 how the limits $pK_{aA}^{*(SH)}$ and $pK_{aB}^{*(SH)}$ evolve when we change the solvent.

3.5. Leveling effect of a solvent

Let us now consider the particular acid–base reactions between an acid AH and the base from an amphoteric protic solvent SH. Any acid that reacts completely with the base SH (the solvent) is called a strong acid. Any base which reacts completely with the acid SH_2^+ is called a strong base.

Now consider the two equilibria between an acid and a base in solution, and the corresponding expressions of the law of mass action. If the solvent is a base, we see the following reaction:

$$SH_2^+ + SH = SH + SH_2^+ \qquad [3R.8]$$

Thus, we have the relation:

$$K_8 = \frac{|H_2S^+|}{|H_2S^+|} = 1 \qquad [3.26]$$

For the acidic nature of the solvent, we have:

$$2SH = S^- + SH_2^+ \qquad [3R.9]$$

The corresponding constant is the ion product of the solvent, so:

$$K_9 = |H_2S^+|\,|S-| \qquad [3.27]$$

We can now plot the diagram of predominance of the different species derived from the solvent (Figure 3.2).

We see that the solvent SH is stable only in the intermediary domain between the points D_1 and D_2 which correspond to equilibria [3R.6] and [3R.8]. That domain is called the *protic activity domain* of the solvent SH.

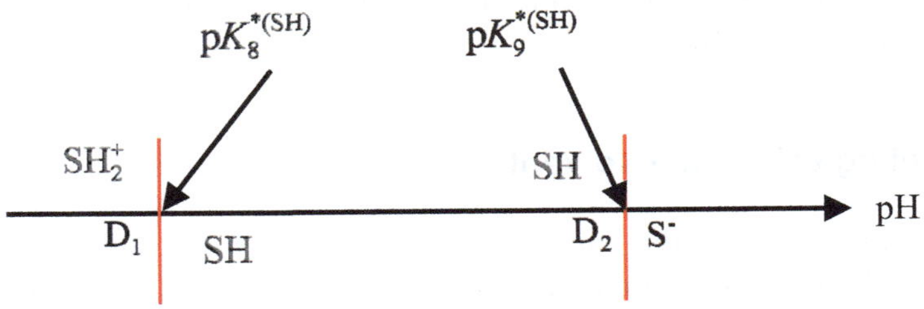

Figure 3.2. *Protic activity domain of an amphoteric solvent*

We now dissolve an acid in that solvent and place the protic activity corresponding to its $pK_{aA}^{*(SH)}$ on the axis in Figure 3.3. We find three scenarios:

– the point which corresponds to the $pK_{aA}^{*(SH)}$ of the acid is to the left of point D_1 (Figure 3.3). In this case, only the base A^- will be stable within the range D_1D_2. The acid will be ionized completely, and is a strong acid;

– the point which corresponds to the point $pK_{aA}^{*(SH)}$ is placed between the points D_1 and D_2 (Figure 3.4). On the left, the acid AH is stable; on the right, it is the base A^- which is stable. We are dealing with a weak acid that is partially dissociated in water;

– the point corresponding to the point $pK_{aA}^{*(SH)}$ is to the right of point D_1 (Figure 3.5). The acid AH will not be dissociated in the solvent and the corresponding base A^- will be a strong base.

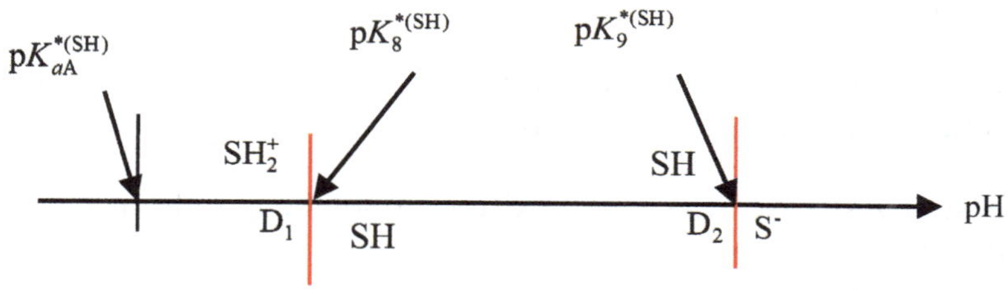

Figure 3.3. *Protic activity domain of a solvent: the case of a strong acid, which is completely ionized*

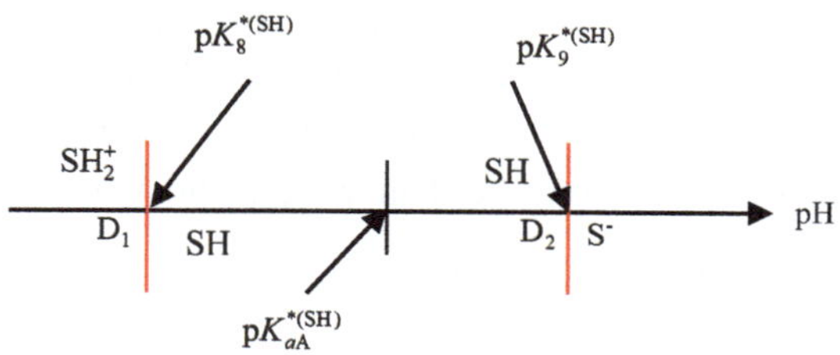

Figure 3.4. *Protic activity domain of a solvent: the case of a weak acid, which is partially ionized*

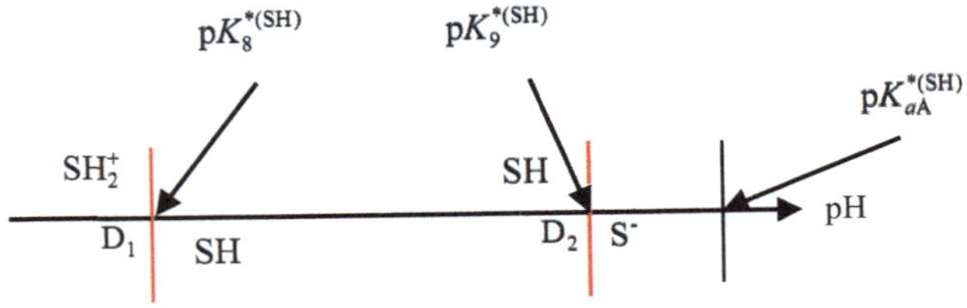

Figure 3.5. *Protic activity domain of a solvent: the case of a non-dissociated acid*

Therefore, we can conclude:

In a solvent SH, any acid stronger than the ion SH_2^+ reacts totally with the solvent.

In a solvent SH, any base stronger than the ion S^- reacts completely to give the solvent.

Thus, the strongest acid that can exist in a solution is the positive ion yielded by the autoprotolysis of the solvent.

The strongest base which can exist in a solution is the negative ion created by the autoprotolysis of the solvent.

This phenomenon is known as the *leveling effect* of a solvent.

The different solvents do not have the same protic activity domain. Figure 3.6 compares the protic activity domains of a number of solvents. The limits of the protic scale are -40 (super-acidic media) and +50 (super-basic media). Thus, not all solvents have the same leveling effect. In Figure 3.6 we see water, whose domain ranges from 0 to 14, where any acid whose $pK_{aA}^{*(SH)}$ is less than 0 is a strong acid in water, and where any acid whose $pK_{aA}^{*(SH)}$ is greater than 0 is a weak acid.

If we compare two solvents (e.g. water and ethanol), an acid AH whose point corresponding to $pK_{aA}^{*(SH)}$ lies between -2 and 0 in the two solvents will be a strong acid in water and a weak acid in ethanol. This is what we call the dissociating effect of solvents. This is what we see, in particular, with hydrochloric acid.

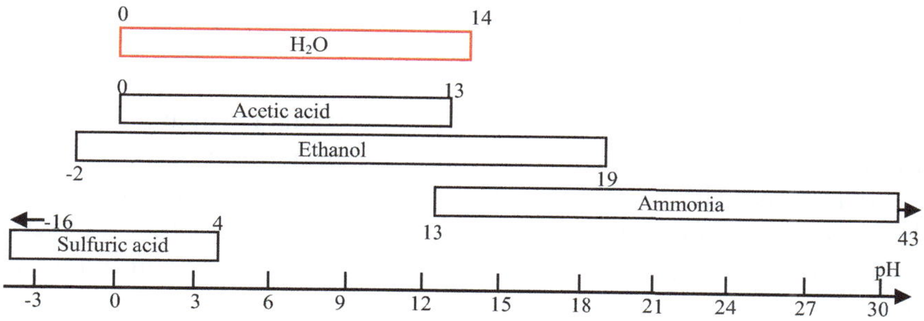

Figure 3.6. *Protic activity domains of a number of solvents*

Hence, we can measure the $pK_{aA}^{*(SH)}$ of hydrochloric acid in ethanol, but it is not possible in water.

Thus far, we have only looked at the effect of amphoteric solvents with a large dielectric constant. For purely acidic or purely basic solvents, obviously, there is only one limit on the protic activity axis – respectively D_1 and D_3.

In solvents with a small dielectric constant, the acidity constant is no longer given by relation [3.17], but by relation [3.15] which, because of the constant K_i, means we cannot have so simple a relation as relation [3.22] to calculate the protic strength in the solvent on the basis of the pH in water.

3.6. Modeling of the strength of an acid

We shall construct a simple model of the acidity constant of an acid in a molecular solvent SH. This model is based on that of the ion pair. We can then use that model to compare two solvents and to plot their respective positions – i.e. construct the diagram shown in Figure 3.6.

3.6.1. *Model of the strength of an acid*

In this model, developed by Bjerrum, an ion pair is modeled as a dipole made up of two punctual charges. The distance between those two charges is chosen as being equal to the sum of the radii of the two ions making up the pair. We denote this distance as a. Its value obviously depends on the acid AH and the solvent SH.

Thus, let us consider a dipole of two ions – carrying the charge +e, for the acid form of the neutral solvent, and $z_B e$, for the basic form of the acid–base pair under study. The charges are a distance a apart, and are placed in a medium with dielectric constant D (that of the solvent). The potential energy of such a dipole, with the origin when the two charges are infinitely far apart, is:

$$E_p = \frac{\mathrm{N_a} e^2 |z_\mathrm{B}|}{4\pi\varepsilon_0 aD} \quad [3.28]$$

Consider the equilibrium of dissociation [3R.5]. Calculate the variation in standard Gibbs energy associated with that equilibrium. This Gibbs energy is the energy needed to separate the ions of the pair, which is given by the opposite of the dipole's potential energy.

Thus, we have:

$$\Delta_D G^0 = -\frac{\mathrm{N_a} e^2 |z|}{4\pi\varepsilon_0 aD} \quad [3.29]$$

Based on this standard Gibbs energy, we can calculate the logarithm of the equilibrium constant pertaining to the concentrations of the dissociation reaction:

$$\log K_D = -\frac{\Delta_D G^0}{2.3\mathrm{R}T} = \frac{\mathrm{N_a} e^2 |z|}{2.3\mathrm{R}T 4\pi\varepsilon_0 aD} \quad [3.30]$$

In 1958, Fuoss [FUO 58], refining Bjerrum's model, calculated the potential energy of the ion pair but, in doing so, took account of the contribution of the ionic atmosphere as calculated by Debye when establishing his ionic solution model. Fuoss found the following expression for the decimal logarithm of the dissociation constant:

$$\log K_D = \frac{\mathrm{N_a} e^2 |z|}{2.3\mathrm{R}T 4\pi\varepsilon_0 aD} + \log\frac{4\pi \mathrm{N_a} a^3 \rho}{3} \quad [3.31]$$

In this expression, ρ is the density of the solvent.

To simplify this expression, we posit:

$$A = \log \frac{4\pi N_a a^3 \rho}{3} \qquad [3.32]$$

Strictly speaking, A depends on the solvent by way of its density. Given that the solvents have fairly similar densities, and even more similar logarithms, we can consider A to be a constant, independent of the solvent and of the acid–base pair under study.

We also posit:

$$\beta = \frac{N_a e^2}{2.3RT4\pi\varepsilon_0} \qquad [3.33]$$

Relation [3.30] is then written thus, moving to pK_D:

$$pK_D = A + \frac{\beta}{a}|z|\frac{1}{D} \qquad [3.34]$$

Taking account of relation [3.14], for pK_a, we find:

$$pK_a = A + \frac{\beta}{a}|z|\frac{1}{D} - \log\alpha_i \qquad [3.35]$$

In this expression, the values a and z depend on the acid and the values a, α_i and D depend on the solvent. A and β are constants.

Ion	Li^+	Cl^-	I^-	OH^-	NH_4^+	H_3O^+	NO_2^-	IO_3^-	NO_3^-	SO_4^{2-}	PO_4^{3-}
Radius (pm)	76	181	220	140	150	150	155	182	189	230	238

Table 3.2. *Ionic radii of a number of ions (in picometers, i.e. 10^{-12} m)*

In practice, the term a, which is the sum of the ionic radii of the basic form of the acid and the acid form of the solvent, has very little dependence on the solvent and solute. This is evident when we look at Table 3.2, which gives the ionic radii of certain ions. A value of 200 pm for a is entirely acceptable. The consequence is that the term β/a is practically constant.

3.6.2. Comparison of an acid's behavior in two solvents

We consider only the case of neutral solvents. The acid forms of these solvents, then, will carry a charge of +1.

Consider two molecular solvents 1 and 2, with respective dielectric constants D_1 and D_2, and respective degrees of ionization α_{i1} and α_{i2}. The charge of the basic form of the acid–base pair is z_B. As noted above, we can suppose that the ratio β/a is constant. We return to relation [3.35] to find the pK_a of the acid AH in each of the two solvents. By finding the difference between the pK_a values, we immediately see that:

$$\Delta(pK_a) = pK_a^2 - pK_a^1 = \frac{\beta|z_B|}{a}\left(\frac{1}{D_2} - \frac{1}{D_1}\right) - \Delta(\log \alpha_i) \quad [3.36]$$

We shall now examine three examples of acids in the solvent SH, and calculate the term $\Delta(pK_a)$, taking the solvent water solvent 2 in the last relation) as a reference. For water, pK_a = 0, as indicated by convention [3.19]. The three acids under examination will differ from one another by the charge z_B of the basic form in the pair AH. As water is being used as the reference point, the calculated difference will be the pK_a of the acid in relation to water – i.e. $pK_{aA}^{*(SH)}$. By this comparison, whatever solvent is used, all acids can be pinpointed on a single scale: the scale of pK_a values in water. The value of the pK_a of the pair H_3O^+/H_2O is taken as the origin of that axis.

3.6.2.1. Acids with a charge of +1

An example of such acids include water (acid H_3O^+), but also ammonium (acid NH_4^+).

The acid-base equilibrium is written as:

$$AH^+ + H_2O = H_3O^+ + A \quad [3R.10]$$

The base carries no charge, so the application of law [3.36] gives us:

$$pK_{aA}^{*(SH)} = -\Delta(\log \alpha_i) \quad [3.37]$$

Thus, $pK_{aA}^{*(SH)}$ is independent of the dielectric constant. Only the difference in basicity of the two solvents comes into play. This result was predictable because the reaction is an isoionic reaction (see section 2.7.2.1), and we ignored the difference between the radii of the ions.

Figure 3.7 represents the $pK_{aA}^{*(SH)}$ of the acid AH as a function of the opposite of the dielectric constant of the solvent. The vertical axis represents the solvent water (dielectric constant around 80). The different solvents are represented by vertical lines. We have illustrated the straight line representing the $NH_3/NH4^+$ pair. The curve, therefore, is a horizontal, shifted in relation to zero. Whatever the solvent, the pair retains the same $pK_{a(NH_4^+)}$, equal to 9.25.

3.6.2.2. *Non-charged acids*

Such is the case, for example, with the ethanoic acid/ethanoate pair (CH_3COOH/CH_3COO^-).

The acid-base equilibrium is written:

$$AH + H_2O = H_3O^+ + A^- \quad [3R.11]$$

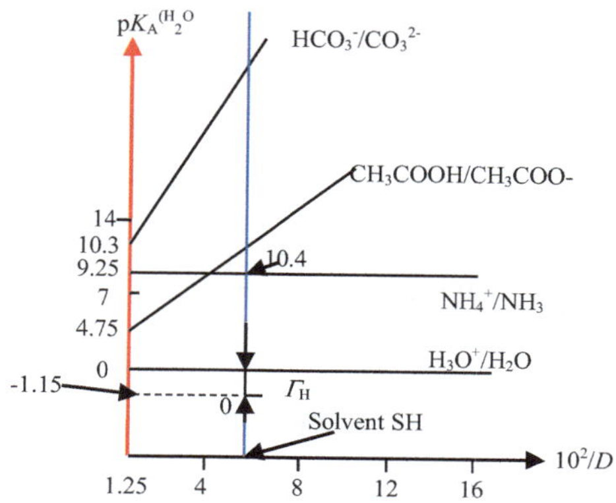

Figure 3.7. *Variations of the acidity constants in relation to water with the constant*

The charge of the base is -1, so the application of the law [3.35] gives us:

$$\mathrm{p}K_{a\mathrm{A}}^{*(\mathrm{SH})} = \frac{\beta}{a}\left(\frac{1}{D_1}\right) - \Delta(\log \alpha_i) - \frac{\beta}{a}\left(\frac{1}{D_2}\right) \qquad [3.38]$$

Hence, $\mathrm{p}K_{a\mathrm{A}}^{*(\mathrm{SH})}$ is a $1/D$ affine function.

In Figure 3.7, the curve is a straight line with the slope β/α, obeying the equation:

$$\mathrm{p}K_{a\mathrm{A}}^{*(\mathrm{SH})} = \frac{\beta}{a}\left(\frac{1}{D}\right) - \Delta(\log \alpha_i) - \frac{\beta}{a}\left(\frac{1}{D_{water}}\right) \qquad [3.39]$$

In Figure 3.7, we have plotted the line representing the ethanoic acid/ethanoate pair, whose $\mathrm{p}K_a$ in water is 4.75. The line cuts the vertical axis at ordinate 4.75.

This case is that of an equilibrium where there is an increase in the number of ions (see section 2.7.2.2).

3.6.2.3. *Acids with a charge of -1*

Such is the case, for instance, with the carbonate/bicarbonate pair (CO_3^{2-}/ HCO_3^-).

The acid–base equilibrium is written:

$$AH + H_2O = H_3O^+ + A^- \qquad [3R.12]$$

The basic charge is -2, so by application of [3.40], we obtain:

$$\mathrm{p}K_{a\mathrm{A}}^{*(\mathrm{SH})} = \frac{2\beta}{a}\left(\frac{1}{D_1}\right) - \Delta(\log \alpha_i) - \frac{\beta}{a}\left(\frac{1}{D_2}\right) \qquad [3.40]$$

Thus, $\mathrm{p}K_{a\mathrm{A}}^{*(\mathrm{SH})}$ is still a $1/D$ affine function.

In Figure 3.7, the curve is a straight line whose slope is $2\beta/\alpha$ (double that of the previous line), obeying the equation:

$$\mathrm{p}K_{a\mathrm{A}}^{*(\mathrm{SH})} = \frac{2\beta}{a}\left(\frac{1}{D}\right) - \Delta(\log\alpha_i) - \frac{\beta}{a}\left(\frac{1}{D_{water}}\right) \qquad [3.41]$$

Figure 3.7 shows the straight line relating to the bicarbonate/carbonate pair, whose $\mathrm{p}K_a$ in water is 10.3. The line intersects the vertical axis at ordinate 10.3.

Again, we are dealing with an equilibrium with an increase in the number of ions (see section 2.7.2.2).

Now, in Figure 3.7, we shall show another solvent SH, represented by a vertical line. Each of the lines, illustrating each acid–base pair, intersects that vertical at a point which represents the $\mathrm{p}K_a^{(\mathrm{SH})}$ of the corresponding pair in that solvent. The difference between that value and the $\mathrm{p}K_{a\mathrm{A}}^{*(\mathrm{SH})}$ of the acid is given by relation [3.19]. This means that the value on the $\mathrm{p}K_a^{(\mathrm{SH})}$ scale on the vertical representing the second solvent can be deduced from the value on the water scale by a translation of $\Gamma_{\mathrm{H}}^{(\mathrm{Tr})}$.

3.6.3. *Construction of activity zones for solvents*

We can now see how it is possible to construct a scale showing the acidity and the relative position of the different solvents – i.e. how to position the different lines for each acid–base pair and how to place the solvents and their limits.

We shall choose the example of ethanol as a solvent. We base our reasoning on Figure 3.8, and proceed in four steps.

3.6.3.1. *Step 1: positioning the solvents*

We plot the abscissa axis, graduated in $10^2/D$ intervals. We position the two vertical axes in their rightful places – one relating to water and the other to ethanol. We graduate the water axis and note on it the domain of protic strength relative to water – i.e. the zone 1-14. On the horizontal, we reproduce this graduation on the $\mathrm{p}K_{a\mathrm{A}}^{*(\mathrm{SH})}$ ordinate axis.

3.6.3.2. *Step 2: calibrating and positioning the solvent scale in relation to the water scale*

The purpose of this step is the place the zero of the scale on any chosen solvent. In order to do so, having positioned the horizontal of the H_3O^+/H_2O pair at zero on the water scale, we plot the horizontal line for an acid–base pair whose pK_a is independent of the dielectric constant. We know that to do this, we have to choose a pair whose base is electrically neutral (see section 3.6.2.1). For instance, we might choose NH_4^+/NH_3. We know that the pK_a of this pair in water is 9.25, whilst its $pK_{aA}^{*(EtH)}$ in ethanol is 10.4. We place both of these values on their correct scale. The difference between these two values (1.15) can be used to position the origin of the ordinate relative to ethanol: it is the point at which the $pK_{aA}^{*(SH)}$ has a value of -1.15. Based on the two ordinate values of 10.40 and 0, we are now able to graduate the ethanol axis.

3.6.3.3. *Step 3: plotting the zone of the protic scale of the solvent*

We now need to delimit the protic scale of ethanol. Suppose we have already measured its ion product, which is:

$$pK_i^{(EtH)} = 19$$

We place that value on the ethanol axis in Figure 3.8. The zone 0-19, therefore, is the zone of protic action of ethanol. Any acid whose $pK_{aA}^{(EtH)}$ lies between these two would be a weak acid, partially ionized in ethanol. Thus, we can define and measure that $pK_{aA}^{(EtH)}$.

3.6.3.4. *Step 4: plotting an acid–base pair with a non-horizontal line*

We now choose an acid–base pair for which the representative line is not horizontal. We need to know its $pK_{aA}^{(H_2O)}$ and its $pK_{aA}^{(EtH)}$. That acid could, for instance, be chosen from among the known electrically-neutral acids (see section 3.6.2.2). For our investigation, let us use ethanoic acid. In water, its $pK_{aA}^{(H_2O)}$ is 4.75; in ethanol, it is $pK_{aA}^{(EtH)} = 10.4$. We place these two values respectively on the water axis and on the ethanol axis in Figure 3.8. We can draw a straight line passing through these two points, which is the line representing ethanoic acid on the graph.

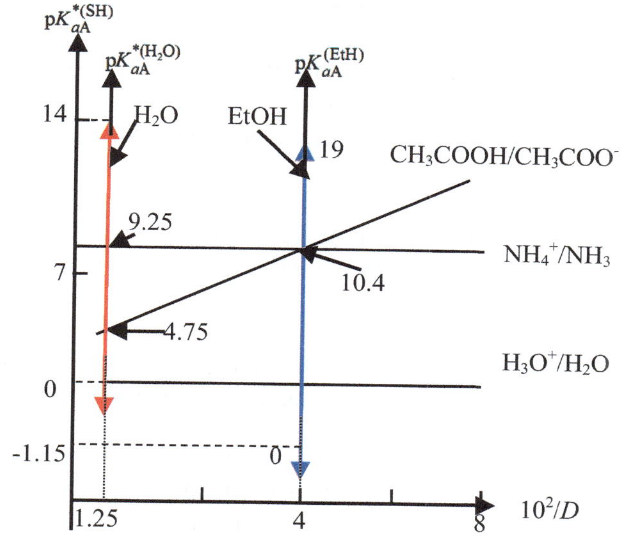

Figure 3.8. *Position of the acidity scales and the different solvents*

Now we can plot all the lines for any given acid–base pair if we simply know its $pK_{aHCl}^{(H_2O)}$ in water or in ethanol. For instance, if we choose to plot the line corresponding to the HCl/Cl$^-$ pair, we are unable to measure the $pK_{aHCl}^{(H_2O)}$ in water because HCl is a strong acid in this pair. On the other hand, in ethanol, we can measure the value of $pK_{aHCl}^{(Eth)}$, which is 2.1. We now plot the line representing HCl, which is parallel to that of acetic acid (see section 3.7.3.2). By extrapolating from that line, we are able to calculate $pK_{aHCl}^{(H_2O)}$, and we obtain around -4, which does indeed show that HCl is a strong acid in water; the representative line intersects the water axis below the value 0.

If we choose to plot the straight line corresponding the HCO_3^-/CO_3^{2-} pair, we know that its $pK_{aHCO_3^-}^{(H_2O)}$ is 10.3. Through that point, we need to plot the representative straight line with a slope twice that of ethanoic acid (see section 3.6.2.3). It is then easy to find $pK_{aHCO_3^-}^{(EtH)}$ by the intersection of the line with the ethanol axis, and we read 19.1, which shows that this value cannot be measured directly because it is situated outside of the zone of protic activity of ethanol.

3.7. Acidity functions and acidity scales

It is often tricky to directly measure the activities of the ions in a non-aqueous medium. Thus, it is worthwhile to find a more practical way to evaluate a solvent's ability to exchange protons with the acid–base pairs dissolved in it.

In addition, we have seen that in water the range of pK_a that it is possible to evaluate runs between 0 and 14, which limits its use to a fairly restrictive range of acid–base pairs.

We are able to solve the two problems thus posed by using more acidic or basic solvents than water and introducing a more extensive acidity scale.

The first people to introduce such relations, in 1932, were Hammet and Deyrup [HAM 32], by what is known as the Hammett acidity function. That function is based on spectroscopic determinations of the medium's acidity. To obtain such a measurement, we place an appropriate colored indicator in the medium. In other words, one of the forms (acid or basic) of that indicator must absorb a sufficiently intensive wavelength so that absorbance measures can be performed in a very dilute medium.

To introduce the Hammett acidity function, consider the equilibrium [3R.13] between a neutral base B_1 and its conjugate strong acid B_1H^+ in an aqueous solution:

$$B_1H^+ + H_2O = B_1 + H_3O^+ \qquad [3R.13]$$

The acidity constant of that pair $K_{BH}{}^+$ is expressed as a function of the activities by:

$$K_{B_1H^+} = \frac{|B_1|\,|H_3O^+|}{|B_1H^+|} = \frac{|B_1|\,a^+}{|B_1H^+|} \qquad [3.42]$$

We can express an activity as the product of the concentration by the corresponding activity coefficient, so the expression of the constant becomes:

$$K_{B_1H^+} = \frac{[B_1]\left[H_3O^+\right]}{\left[B_1H^+\right]}\frac{\gamma_{B_1}\,\gamma_{H_3O^+}}{\gamma_{B_1H^+}} \qquad [3.43]$$

Using a sufficiently dilute solution so that the activity coefficients can be taken as equal to 1 (which may not necessarily be the case in solution in a different solvent), relation [3.43] becomes:

$$K_{B_1H^+} = \frac{[B_1]\left[H_3O^+\right]}{\left[B_1H^+\right]} \qquad [3.44]$$

From this relation, we deduce:

$$pK_{B_1H^+} = pH - \log\frac{[B_1]}{\left[B_1H^+\right]} \qquad [3.45]$$

If the base B_1 (or its conjugate acid) is a colored indicator, the ratio of concentrations $[B_1]/[B_1H^+]$ can be measured experimentally by light absorption.

The value of that ratio is a measure of the solution's tendency to transfer a proton to a neutral base.

By measuring the pH, which presents no difficulty at all in a dilute aqueous medium, we can very easily deduce the value of $pK_{BH}{}^+$.

Now consider another neutral base B_2 such that, in a given solvent, its protic activity is a+. Suppose that its acidity constant in relation to water is lower than that of the previous pair. For this new pair, we can write:

$$pK^{*(SH)}_{B_2H^+} = \frac{[B_2]\,a^+}{\left[B_2H^+\right]}\frac{\gamma_{B_2}}{\gamma_{B_2H^+}} \qquad [3.46]$$

We can then use a solvent SH such as a mixture of water and 10% H_2SO_4 to measure both the ionization of B_1 and that of B_3. In this new solvent, according to equations [3.45] and [3.46], we would have:

$$pK^{*(SH)}_{B_2H^+} - pK^{*(SH)}_{B_1H^+} = -\log\frac{[B_2]}{\left[B_2H^+\right]} + \log\frac{[B_1]}{\left[B_1H^+\right]} + \log\frac{\gamma_{B_1}\gamma_{B_2H^+}}{\gamma_{B_2}\gamma_{B_1H^+}} \qquad [3.47]$$

If the bases B_1 and B_2 are the same type, it is highly likely that in a medium of average ionic strength, the third term in equation [3.47] will be practically the same no matter what the solvent. As that term is equal to zero in a dilute aqueous solution, it is also zero in other solvents. This is illustrated by Figure 3.9. For different bases of the same type, we have plotted $\log([B]/[BH^+])$ as a function of the acid content of a solvent of a mixture of water and sulfuric acid. The lines representing each base are all parallel to one another.

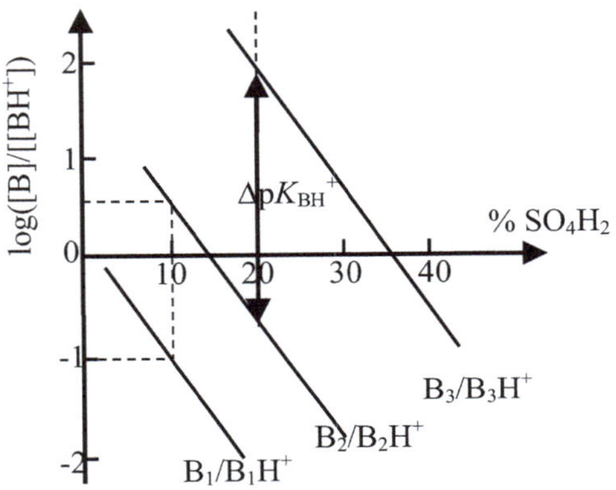

Figure 3.9. *Logarithm of the ratio between the concentrations of the base and of its conjugate acid, for different bases of the same type, depending on the solvent's composition*

Hammett defines the *acidity function* H_0 of an acid solution in relation to a neutral base by:

$$H_0 = \mathrm{p}K_{\mathrm{BH^+}}^{*(SH)} + \log\frac{\left[\mathrm{BH^+}\right]}{\left[\mathrm{B}\right]} = -\log\frac{\gamma_{\mathrm{B}} a^+}{\gamma_{\mathrm{BH^+}}} \qquad [3.48]$$

The function H_0 of a solution determines the ratio of the activities $|B|/|BH^+|$. It quantifies the tendency of that solution to cede protons to a neutral base. It indicates the pH, to which it is identical in an aqueous solution. Because the ratio between the activity coefficients $\frac{\gamma_{\mathrm{B}}}{\gamma_{\mathrm{BH^+}}}$ is independent of the base, H_0 is, like the protic activity a^+, a characteristic property of the medium.

We can see that this function is only truly useful if it is independent of the base B – i.e. when the ratio γ_B/γ_{BH+} is the same for different bases B. This is the situation we observe with similar bases, which might only differ from one another, say, by the substitution of one chemical group for another. Table 3.3 gives a few indicators for a family of nitranilines and their pK values, which have been used to determine values of H_0.

Thus, for a family such as this, we simply need to know K_B in a dilute solution and the ratio of the concentrations $[BH_+]/[B]$, determined using spectrography in the concentrated solution, to obtain the function H_0.

Base	p-nitraniline	o-nitraniline	p-nitrodiphenylamine
$pK_{BH^+}^{(H_2O)}$	1.11	-0.13	-2.98
Base	2-4-dinitraniline	6-bromo-2,4-dinitraniline	2, 4, 6-trinitraniline
$pK_{BH^+}^{(H_2O)}$	-4.38	-6.59	-9.29

Table 3.3. *The pK values of some anilines substituted in water*

This original list of indicators has expanded significantly, because we can use NMR dosing of the proton instead of using absorption spectroscopy.

Table 3.4 shows a selection of values of the function H_0 in water–sulfuric acid mixtures.

% SO_4H_2	5	10	20	30	40	50	60	70	80	90	100
$-H_0$	-0.24	0.16	0.89	1.54	2.28	3.23	4.32	5.54	6.82	8.17	10.60

Table 3.4. *Values of the acidity function in mixtures of water and sulfuric acid*

Note that two media with the same value H_0 do not necessarily have the same value of the protic activity a^+. Indeed, if we look at the second part of relation [3.48], in order to have the same value of a^+ at the same value H_0, the ratios of the activity coefficients of the base need to be identical; however, they may vary considerably from one medium to another.

NOTE.– Other acidity scales have been defined on the basis of different families of bases. Thus, with a family of bases, charged negatively, the function H_- has been defined. Of course, the values of H_0 and H_- do not match, because the ratios between the activity coefficients are not the same. The function H_0 is used in the domains where the values of H are negative; the function H_-, which again uses nitroanilines but in acid form, is used for values of H higher than 14. Similarly, we define a function H_+ which measures the solvent's tendency to transfer a proton to a cationic base.

One of the major qualities of the functions H is that they do not depend on the solvent – at least those which are sufficiently ionizing.

It is clear, as we have seen, that the function H and the pH are the same in highly dilute aqueous solutions. This is no longer true in concentrated solutions. Figure 3.10 shows the comparison between the value H_0 and that of the pH in solutions of sulfuric acid in water.

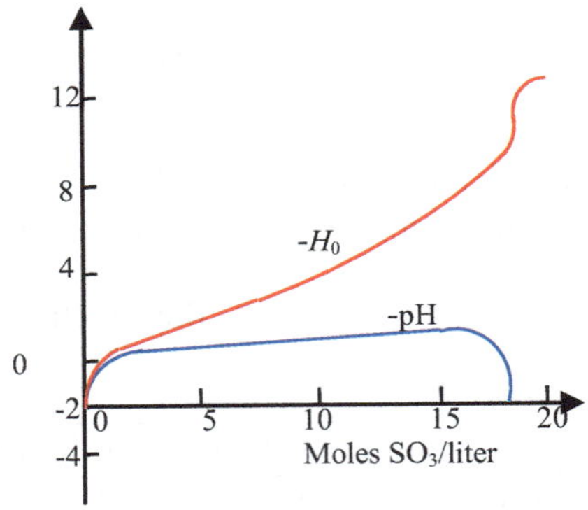

Figure 3.10. *Comparison of the values of H_0 and pH in solutions of sulfuric acid in water*

We can see that, in water, the amplitude of the variation of the function H_0 is significantly greater than that of the pH.

3.8. Applications of the acidity function

We shall now give a few applications of the function H_0, which can be used to evaluate certain constants without employing electrochemical measuring techniques, which are not always easy to use in non-aqueous media.

3.8.1. *Measuring the pKa of an indicator*

Consider an indicator B_3. We want to determine the value of $pK^{*(SH)}_{B_3H^+}$. We place it in a solvent so that it also covers the domain of a solvent B_2 with known $pK^{*(SH)}_{B_2H^+}$. We measure the value of $\log([B_3]/[B_3H^+])$. Let us plot this value in Figure 3.9 at the ordinate relative to the solvent used. We determine the point H_3. We can write, between the point H_3 and the point H_2, situated on the same vertical, that the two media have the same value of the Hammett acidity function – i.e.:

$$\Delta H_0 = 0 \qquad [3.49]$$

From this, we deduce the relation:

$$pK^{*(SH)}_{B_3H^+} - pK^{*(SH)}_{B_2H^+} = \Delta\left(pK^{*(SH)}\right) = -\log\frac{[B_3]}{\left[B_3H^+\right]} + \log\frac{[B_2]}{\left[B_2H^+\right]} \qquad [3.50]$$

We can measure the two logarithmic terms by absorption spectroscopy, which gives us the difference $\Delta\left(pK^{*(SH)}\right)$, and thus we can calculate $pK^{*(SH)}_{B_3H^+}$.

The curve passing through H and relative to B_3 is a parallel to the curve relative to the indicator B_3. The difference $\Delta\left(pK^{*(SH)}\right)$ is the same no matter what the solvent. Thus, we can plot that line and thus determine $pK^{*(SH)}_{B_3H^+}$ in any given solvent if we know the line relative to B_3.

3.8.2. *Measuring the ion products of solvents*

Consider a solvent SH . The autoprotolysis reaction is written:

$$2SH = SH_2^+ + S^- \qquad [3R.14]$$

The ionization constant, then, is defined by:

$$K_{SH} = |S^-|\,|SH_2^+| \qquad [3.51]$$

We want to measure K_{SH}. In order to do so, we dissolve a strong base at concentration C in that solvent. Thus, we have:

$$|SH_2^+| = \frac{K_{SH}}{C} \qquad [3.52]$$

Let us also dissolve an indicator B, which is appropriate for that solvent, at total concentration C_B. By Raman spectroscopy and that value of C_B, we determine the ratio [B]/[BH$^+$]. Thus, we calculate H_0 on the basis of the relation [3.48]. Based on the same relation, we can write:

$$H_0 - \log\left[SH_2^+\right] = pK_B^{*(SH)} + \log\frac{[B]\left[SH_2^+\right]}{BH^+} = pK_B^{*(SH)} - pK_{C(B)}^{(SH)} \qquad [3.53]$$

Suppose that the indicator is chosen such that we have the following approximation:

$$\frac{[B]\left[SH_2^+\right]}{BH^+} \cong -pK_{(B)}^{(SH)} \qquad [3.54]$$

Relation [3.53] becomes:

$$H_0 = \log\left[SH_2^+\right] + pK_B^{*(SH)} - pK_{(B)}^{(SH)} \qquad [3.55]$$

In view of relation [3.52], relation [3.55] becomes:

$$H_0 = pK_B^{*(SH)} - pK_{(B)}^{(SH)} - \log K_{SH} + \log C \qquad [3.56]$$

Thus if approximation [3.54] is acceptable, if we plot the curve representing the function $H_0 = f(\log C)$, we should obtain a straight line whose slope is $pK_B^{*(SH)} - pK_{(B)}^{(SH)} - \log K_{SH}$. If the indicator is known, then from that slope we can deduce the ionization constant of the desired solvent.

If pK_{SH} is known, the position of the solvent SH in Figure 3.7 is known, and its protic activity ranges from 0 to pK_{SH}. The neutrality point in that solvent then has the value $pK_{SH}/2$.

We shall see another method for determining the ion product of a solvent by potentiometry (section 7.7.6).

3.9. Acidity in non-protic molecular solvents

A non-protic solvent – i.e. one whose molecule does not contain a hydrogen atom liable to be ceded in the form of a proton – obviously cannot cede any protons. It has no acidic nature in the sense that Brønsted intended. Therefore, it cannot give rise to an equilibrium of autoprotolysis – that is to say, it cannot have an ion product. We distinguish between two cases depending on whether or not the solvent has a protophilic nature.

If the solvent is protophilic in nature, it can capture protons from an acid, but it is impossible for it to cede a proton to a base. Thus, acids dissolved in such solvents are more or less dissociated, whilst bases exhibit no dissociation whatsoever. We can find more or less strong acids in these solvents. The strongest acid which can exist is SH^+, which results from the fixation of a proton on the molecule of the solvent S.

Each acid–base pair dissolved in that solvent will, if the acid is weak, exhibit an acidity constant and therefore a pK_a. The acidity scale which obviously starts with the SH^+ is, in principle, limitless in terms of higher pH values. There is no neutrality.

The pH in such a solvent, if it does not contain any acid (pure solvent or solutions of pure bases), remains undetermined.

Of these non-protic but protophilic solvents, we can cite dioxane (C_2H_4O), tetrahydrofuran (C_4H_8O), DMF (dimethylformamide, $CHONH_2$), etc.

If the solvent has no protophilic nature at all, it does not capture the proton, and the acids dissolved in it are not dissociated. There is no longer an acidity constant.

This property does not impede acid–base reactions between solutes, which exchange protons with one another without the intervention of the solvent. Of course, these equilibria are characterized by a law of mass action and an equilibrium constant. It is always possible to classify these equilibria in the same solvent on the basis of the pK_a values of these pairs.

3.10. Protolysis in ionic solvents (molten salts)

In molten salts, we can dissolve acids and bases which are ionic or molecular, and therefore carry out proton exchanges in this medium. The ions making up the solvent may themselves take part in these exchanges, and, as is the case with molecular solvents, we find ionic solvents that are amphoteric or only protophilic.

We can distinguish between two cases of amphoteric ionic solvents:

One of the ions making up the ionic solvent is itself amphoteric. An important family of these solvents is molten alkaline hydroxides. The hydroxide anion is both protophilic and protogenic, and therefore lends the solvent its amphoteric nature. The equilibrium of autoprotolysis is written:

$$2HO^- = H_2O + O^{2-} \qquad [3R.15]$$

Thus, in this medium, dissolved water constitutes the strongest acid. Conversely, the oxide ion O^{2-} constitutes the strongest base. Numerous dissolved acids are strong in this medium – e.g. phenol, which dissolves as follows:

$$C_6H_5OH + HO^- = H_2O + C_6H_5O^- \qquad [3R.16]$$

The hydride ion is a strong base in these molten alkaline hydroxides. It reacts as follows:

$$H^- + HO^- = O^{2-} + H_2 \qquad [3R.17]$$

The measure of acidity, equivalent to the pH in aqueous solutions, is the pH_2O.

In another family of ionized solvents, the amphoteric nature of the solvent is due to one of the ions, which is protogenic, whilst the other is protophilic. As an example of such a solvent, we could cite molten ethylammonium chloride (melting point 108°C). Here, hydrochloric acid is the strongest acid, whilst ethylamine, which is the conjugate base of the ethylammonium cation, is the strongest base. Numerous bases are strong when dissolved in this solvent. Examples include the oxide ion, the hydroxide ion, the ethanoate ion and ammonia. The extent of the acidity range is then measured by a pHCl scale, which extends for around 8 units. The equilibrium of autoprotolysis is written thus:

$$C_2H_5NH_3^+ + Cl^- = C_2H_5NH_2 + HCl \qquad [3R.18]$$

NOTE.– In ionic solvents, the activities of the ions making up the solvent are taken to be equal to 1.

3.11. Other ionic exchanges in solution

The definition of acids, according to Brønsted, is based on the set of reactions of proton exchanges. The development of that domain of reactions in solution is due essentially to the fact that these reactions lead to chemical equilibria. This facilitated the introduction of the concepts of pK_a, of domains of predominance and pH. Specialists in the field of reactions in solution wondered whether it would not be possible to harmonize these concepts in the context of other reactions in solution, but characterized by the exchange of chemical particles other than the proton. This set of coherent approaches to such reactions is known as *ionoscopy*. The particular case where the particles exchanged are protons is *protonotropy*.

3.11.1. *Ionoscopy*

It was Gutmann and Lindqvist who, in the 1950s, developed this new unifying concept of ionoscopy. The notion of a *donor–acceptor pair* was thus introduced. Couples of chemical substances exchange various simple ions such as the halide ion, an oxide ion O^{2-}, a sulfide ion S^{2-}, etc. Thus, in each couple, we have a donor and a conjugate acceptor of the particle being exchanged. These exchange reactions lead to equilibria, which again enables us to use concepts introduced into the study of Brønstedian acid–base reactions: equilibrium constants, domains of predominance and the equivalent of the pH, which would be the pX if the ion exchanged is a halide ion X^-, the pO^{2-} if the ion exchanged is an oxide ion, etc.

NOTE.– In the next chapter, we shall see another type of particle exchange. This time, it is electrons that are being exchanged, rather than chemical particles, and this will constitute the domain of redox reactions (see section 4.2).

Ionoscopy developed in the particular case of solutions in a molten-salts medium with exchange of O^{2-} ions. This family of reactions is dubbed *oxo-acidity* (see section 3.11.2).

As we saw in the case of the proton, the solvent itself is able to participate in these new exchanges of chemical particles. It can be a particle acceptor or donor, or even both at once. Therefore, it is amphoteric.

For example, liquid sulfur dioxide is a solvent which can act as an acceptor of the chlorine ion:

$$SO_2 \text{ (acceptor)} + (C_6H_5)_3CCl \text{ (donor)} = (C_6H_5)_3C^+ + SO_2Cl^- \qquad [3R.19]$$

3.11.2. *Acidity in molten salts: definition given by Lux and Flood*

An ion that is very frequently exchanged in molten salts is the O^{2-} ion. This family of reactions, termed *oxidotropy*, led to a new concept in acidity: *oxo-acidity*. Flood put forward a new definition of acids and bases in molten salts, based on that exchange of an oxygen ion. In order to avoid confusion with conventional acids and bases based on the exchange of protons, it was proposed to use the term "*oxo-acids*" to denote acceptors of oxygen ions and "*oxo-bases*" for donors of that ion. The oxo-acid–oxo-base couple thus works on the equilibrium:

$$\text{Oxo-acid (acceptor)} + O^{2-} = \text{Oxo-base (donor)} \qquad [3R.20]$$

As an example, we can cite uranium trioxide, which is an oxo-base, according to:

$$UO_2^{2+} + O^{2-} = UO_3 \qquad [3R.21]$$

Certain compounds, such as water, are amphoteric, and work equally well as an oxo-base and as an oxo-acid, respectively through the two reactions:

$$2H^+ + O^{2-} = H_2O \quad [3R.22]$$

$$H_2O + O^{2-} = 2OH^- \quad [3R.23]$$

NOTE.– In order to take account of the fact that the charge of the particle exchanged is negative instead of positive (as is the case of a proton), it is the acceptor that is the oxo-acid, rather than the donor.

Donor–acceptor couples are characterized by an acidity constant, which is written as follows for the reaction [3R. 20]:

$$K_a = \frac{|\text{acceptor}|}{|\text{donor}|\,|O^{2-}|} \quad [3.57]$$

Note that, as with Brønsted's acids, the higher the value of K_a, the smaller that of pKa, and the stronger the acid is.

Similarly to a reaction with proton exchange, we would have:

$$pO^{2-} = -pK_a - \log\frac{|\text{acceptor}|}{|\text{donor}|} \quad [3.58]$$

If the solvent is itself an acceptor (oxygenated solvent) and donor, the pure solvent presents a point of neutrality, and the different couples can be classified into two categories: strong acids and weak acids, depending on the value of pK_a in relation to the domain of pO^{2-} of the solvent.

In the case of non-oxygenated solvents, we cannot define a pO^{2-} of the pure solvent, and there is no point of neutrality. As the solvent is not involved in the reaction, we content ourselves with classifying couples by their pK_a. Figure 3.11 gives a classification of different couples in the solvent LiCl-KCl at 450°C.

We can see on this scale that the strongest acid is PO_3^- and the strongest base is CO_3^{2-}.

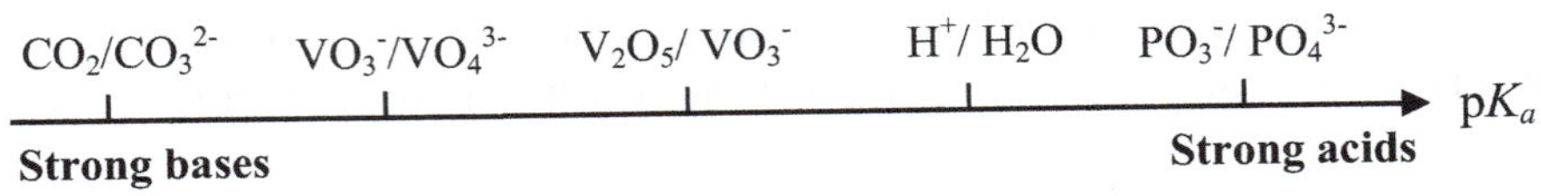

Figure 3.11. *Scale of oxo-acid/oxo-base pairs (taken from Guetta, Fouletier and Taxil [GUE 09]*

3.12. Franklin and Gutmann's solvo-acidity and solvo-basicity

The above definitions of acids and bases were founded on the nature of the ionic particle exchanged – H^+ or O^{2-} – independently of the solvent. However, Franklin and Gutmann put forward a definition of acidity known as *solvo-acidity*.

3.12.1. *Definition of solvo-acidity*

Solvo-acidity is defined on the basis of solvents which are both donors and acceptors of an ionic particle. Such solvents play host to a reaction known as auto-ionization (or auto-dissociation to include ionic solvents), which defines which ion is exchanged during the course of that reaction. That ion therefore helps to define the nature of the pairs in question, known as solvo-acid–solvo-base pairs.

We have already seen the particular case of Brønsted acids and bases in protic amphoteric solvents such as water, where the ion exchanged was the proton. Similarly, in oxo-acidity, oxygenated molten salts (hydroxides, sulfates, carbonates) are amphoteric solvents, whose auto-dissociation results in O^{2-} as the exchanged ion.

3.12.2. *Solvo-acidity in molecular solvents*

Numerous molecular solvents exhibit a reaction of slight auto-ionization, which means they can always be considered to be polar solvents. Therefore, a wide variety of ions can be exchanged in these situations. As examples we could cite halide ions, protons of course, but also the oxygen ion O^{2-}. For example, phosphorus trichloride undergoes auto-ionization as follows:

$$2POCl_3 = POCl_2 + Cl^-, POCl_3 \quad [3R.24]$$

Similarly, bromine fluoride undergoes auto-ionization as follows:

$$2BrF_3 = BrF_2^+ + BrF_4^- \qquad [3R.25]$$

The equilibrium of auto-ionization of a solvent defines the system of the solvent. It involves two characteristic ions: one cation and one anion.

In that system, a solvo-acid is a solute which plays the role of donor of the cation characteristic of the solvent by direct dissociation, or that of acceptor of the characteristic anion in the reaction with the solvent.

A solvo-base is a solute playing the role of donor of the characteristic ion which can be either an anion or a cation.

With this definition, in amphyprotic solvents SH, which are included, the acceptors in S^- act as solvo-acids.

As the system of the solvent defines two characteristic ions, it is agreed to specify the nature of the solvo-acidity determined by the system of the solvent. Thus, we use the term "chloro-acidity" to denote solvo-acidity in solvents such as $POCl_3$, in which the ion exchanged is the chlorine ion.

Thus, consider a solvent whose molecule, represented by SCl, can play the role either of donor or acceptor of the chlorine ion. In the pure state, this solvent is the site of an equilibrium of auto-ionization which, attended by the solvation of the ion exchanged, would be written:

$$2SCl = S^+ + SCl_2^- \qquad [3R.26]$$

As the progress rate of this reaction at equilibrium remains very low, we can define its equilibrium constant by an ion product (the activity of the solvent remains equal to 1):

$$K_{i(S)} = \left|S^+\right| \left|SCl_2^-\right| \qquad [3.59]$$

That equilibrium is affected by the dissolution of chloro-acids (molecules accepting a chlorine ion) or chloro-bases (molecules donating a chlorine ion).

For example, sulfur trioxide is a chloro-acid because of the formation of the chloro-sulfate ion, as follows:

$$SO_3 + SCl = SO_3Cl_2^- + S^+ \qquad [3R.27]$$

Depending on the degree of ionization of the solutes, we may be dealing with strong chloro-acids, which dissociate completely in the solvent, or weak chloro-acids, which dissociate only partially. Similarly, we have strong chloro-bases and weak chloro-bases.

The characteristic value of the medium, then, is pCl^-, which is defined by:

$$pCl^- = -\log\left|SCl_2^-\right| \qquad [3.60]$$

The $pK_{i(S)}$ of the solvent defines the pCl^- scale, in which the solvent is active and the neutrality point equal to $pK_{i(S)}/2$.

Thus, we find all of the reasoning we discussed above in the case of protic solvents. The transposition becomes easy.

NOTE.– As we already saw with Flood's oxo-acidity, because the chlorine ion is an anion, it is chloro-bases which act as donors.

3.12.3. *Solvo-acidity in molten salts*

In ionic solvents which are constituted by molten salts, we see examples of chloro-acidity, such as in sodium tetrachloroaluminate. In molten salts, we can cite a particularly important case on the industrial level: *fluoro-acidity* in cryolith ($NaAlF_6$), which is used in the manufacture of aluminum by electrolysis.

Cryolith dissociates in accordance with the following pattern:

$$NaAlF_6 = AlF_6^{3-} + 3Na^+ \qquad [3R.28]$$

$$AlF_6^{3-} = AlF_5^{2-} + F^- \qquad [3R.29]$$

$$AlF_5^{2-} = AlF_4^- + F^- \qquad [3R.30]$$

Reaction [3R.28] is total to the right, and therefore that dissociation involves two equilibrium constants:

$$K_{29} = \frac{|F^-|\,|AlF_5^{2-}|}{|AlF_6^{3-}|} = 0,25 \qquad [3.61a]$$

$$K_{30} = \frac{|F^-|\,|AlF_4^-|}{|AlF2_5^{3-}|} = 0,05 \qquad [3.61b]$$

It has been shown that in the usage conditions, it is the ion AlF_5^{2-} which is in the majority. In pure cryolith, we have $pF^- = 0.37$.

In that solvent, we dissolve the fluoro-base AlF3 – in practice at content levels between 0 and 0.35. Figure 3.12 shows the scale of pF^- converted depending on the AlF_3 content.

Figure 3.12. *Position on the pF-scale of solutions in cryolith used in electrolysis of aluminum*

Another important family that we encountered is that of oxo-acidity in oxygenated solvents. For example, in molten alkaline carbonates, the reaction of dissociation of the solvent is:

$$CO_3^{2-} = CO_2 + O^{2-} \qquad [3R.31]$$

Thus, CO_2 is the strongest oxo-acid, and hence the boundary of the pO^{2-} scale is the value obtained in the presence of carbon dioxide at the highest partial pressures. The ion product of the solvent takes the form:

$$K_{i(CO3^{2-})} = P_{CO_2}\left[O^{2-}\right] \qquad [3.62]$$

O^{2-} acceptors stronger than CO_2 will be completely dissociated. Such is the case, for instance, with SO_3.

3.13. Acidity as understood by Lewis

We cannot conclude this chapter on acid–base reactions without mentioning the definition of acids and bases put forward by Lewis. Although this new nomenclature no longer pertains only to exchange reactions, it is very useful – particularly in the field of organic chemistry.

Remember that for Lewis, an acid is a compound which accepts a doublet of electrons and a base is a compound which has an available doublet of electrons.

This definition also brings into play the reactions of formation of complex ions, which we shall see in the next chapter (see section 4.1), in the field of acid–base reactions.

4

Complexations and Redox Equilibria

4.1. Complexation reactions

Let us begin by noting that a *complex ion* is a construction formed of an elementary central ion (usually a metal cation), known as the *coordinating element* (or *nucleus*), surrounded by a certain number of molecules – usually simple (water, ammonia, etc.) – which are bonded to it. These molecules are known as the *coordinated groups* or *ligands*. The number of ligands bound to the central atom is called the *coordination number*.

We distinguish between complexes having only one species of ligand (denoted as ML_n) and complexes having ligands of different types (written as $ML_nL'_{n'}$). For example, for complexes with a single species of ligand, we could point to the ion $[Co(NH_3)_6]^{3+}$, which is the luteocobaltic complex; for complexes with several ligands of different types, the praseocobaltic ion $[Co(NH_3)_4Cl_2]^+$ offers a good example.

4.1.1. *Stability of complexes*

If we consider the equilibrium of formation of a complex from its constituent elements, we define the *association constant*, which would be the constant of the equilibrium of formation. Thus, for the complex ML_n, the reaction of formation is:

$$M + nL = ML_n \qquad [4R.1]$$

Thus, the association constant would be:

$$K_a = \frac{[\mathrm{ML}_n]}{[\mathrm{M}]\,[\mathrm{L}]^n} \quad [4.1]$$

The *dissociation constant* would be the equilibrium constant of the inverse reaction – i.e.:

$$\mathrm{ML}_n = \mathrm{M} + n\mathrm{L} \quad [4R.2]$$

Thus:

$$K_d = \frac{[\mathrm{M}]\,[\mathrm{L}]^n}{[\mathrm{ML}_n]} \quad [4.2]$$

Obviously, we would have:

$$K_d K_a = 1 \quad [4.3]$$

As dissociation increases the number of particles which are present, the dilution of the species increases when dissociation takes place.

Thus, a complex is defined by two values: the dissociation constant K_d (or the association constant) and the coordination number n.

Sometimes, with the same cation and the same ligand, there are intermediary complexes which exist between the central ion and the final complex ML_n. Such complexes are denoted as ML_{n-1}, ML_{n-2}, etc.

For each of them, obviously, we define a dissociation constant such that:

$$K_{d_{n-1}} = \frac{[\mathrm{ML}_{n\text{-}1}]\,[\mathrm{L}]}{[\mathrm{ML}_n]} \quad [4.4]$$

For any given series, it is easy to show that the overall dissociation constant is the product of the intermediary dissociation constants:

$$K_d = \prod_{i=n-1}^{i=1} K_{d_{n-i}} \quad [4.5]$$

Complexes are often classified into two categories: perfect and imperfect complexes.

A complex is said to be *perfect* if its dissociation constant is very small. The result of this is that the central ion no longer reacts with its own reagents, because it is practically never found in solution.

A complex is said to be *imperfect* if its dissociation constant is very large.

However, this distinction between perfect and imperfect complexes is imprecise, because how are we to label complexes whose dissociation constant has an intermediary value? We say that a complex is perfect, in the presence of a given reagent, if that complex does not react with that reagent. Thus, a given complex may exhibit perfect behavior in the presence of such a reagent, but imperfect behavior in the presence of another. Ultimately, it depends both on the dissociation constant and the speed of that dissociation, which is obviously quicker if a reagent for the cation M is added and shifts the equilibrium of dissociation to the right.

As we have the concept of the dissociation constant at our disposal, we can also define the *domains of predominance* of the different species.

Let us first consider a complex with a single coordination number ML_n. Finally, we calculate:

$$pL = \frac{pK_d}{n} + \ln\frac{[M]}{[ML_n]} \quad [4.6]$$

Thus, on an increasing pL scale, it is easy to pinpoint the respective domains of predominance of the complex and of the central ion M (as illustrated by Figure 4.1).

if $pL < pK_d$, the form ML_n is predominant;

if $pL > pK_d$, the form M is predominant.

If the complex has intermediary species – e.g. ML, ML_2 and ML_3 – then for each class, in order, we would have the dissociation constants and, on the pL scale, the distribution of the different species between the different pK_{di} values.

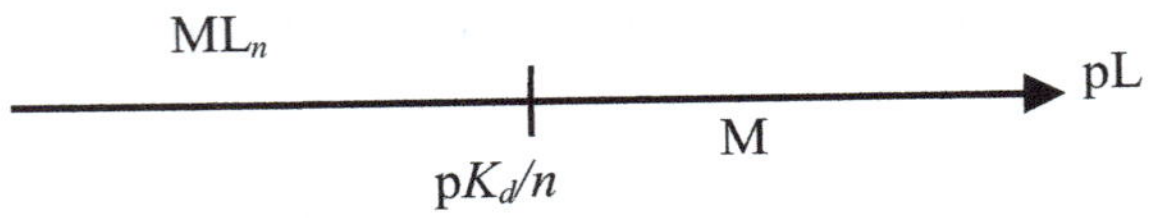

Figure 4.1. *Domains of predominance of the species in a complex MLn*

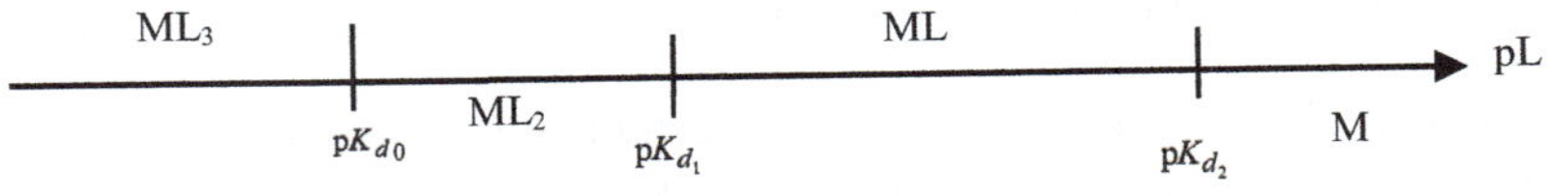

Figure 4.2. *Domains of predominance for a complex with three compositions*

Figure 4.2 illustrates the domains of predominance in the case of three successive complexes: ML_3, ML_2 and ML.

It is also possible to visualize the fractions of the different species at equilibrium, as a function of the concentration of ligand, using a distribution diagram. The fraction of each species present is defined by:

$$\alpha_i = \frac{[\mathrm{ML}_i]}{\sum_{i=0}^{n}[\mathrm{ML}_i]} = \frac{[\mathrm{ML}_i]}{[\mathrm{M}]_{tot}} \quad [4.7]$$

Let K_i denote the overall constant of formation of the complex ML_i from the elements M and L. We can then write:

$$[\mathrm{ML}_i] = K_i[\mathrm{M}]\,[\mathrm{L}]^i \quad [4.8]$$

and:

$$[\mathrm{M}]_{tot} = \sum_{i=0}^{n}[\mathrm{ML}_i] = \sum_{i=0}^{n} K_i[\mathrm{M}]\,[\mathrm{L}]^i \quad [4.9]$$

By feeding expressions [4.7] and [4.8] back into relation [4.9], we obtain:

$$\alpha_i = \frac{K_i[\mathrm{L}]^i}{\sum_{i=0}^{n} K_i[\mathrm{L}]^i} \quad [4.10]$$

and of course:

$$\sum_{i=0}^{n} \alpha_i = 1 \qquad [4.11]$$

The distribution fractions depend only on the overall constants of formation and the concentration of ligand; they are independent of the concentration of the central atom [M] (see relation [4.10]). Curves showing the distribution fractions as a function of the ligand concentration are called *distribution diagrams*. Figure 4.3 shows the example of such a diagram in the case of a series of four complexes with the same central atom bound to four different quantities of ligand (ML, ML_2, ML_3 and ML_4).

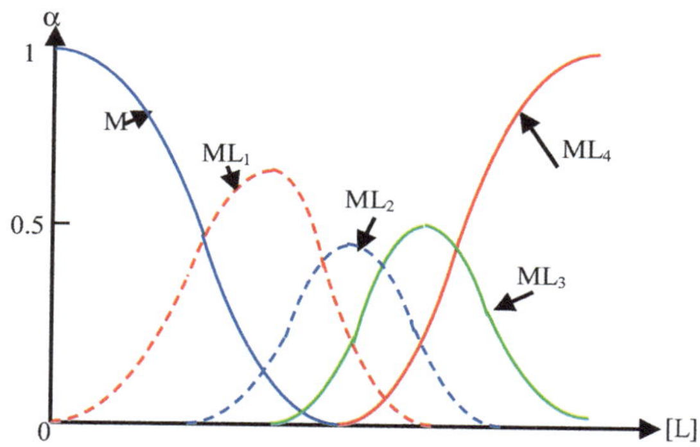

Figure 4.3. *Distribution fractions of complexes with a varied number of ligands. For a color version of the figure, see www.iste.co.uk/soustelle/equilibria.zip*

The dissociation of a complex is obviously influenced by the pH of the solution whenever the ligands are capable of reacting with water.

Take, for example, the complex formed by iron with fluorine FeF^{2+}. This complex dissociates as follows:

$$FeF^{2+} = Fe^{3+} + F^- \qquad [4R.3]$$

The F^- ion, which is a weak base, is liable to react with water as follows:

$$F^- + H_2O = HF + OH^- \qquad [4R.4]$$

Thus, this latter equilibrium is superposed on the former, and means that the dissociation will be influenced by the pH of the solution. A high pH, meaning an alkaline (basic) medium, will drive the two reactions back towards the left, and thus discourage the decomposition of the complex.

Figure 4.4, constructed using the acidity constant K_a of hydrofluoric acid, confirms the expected result, showing that the domain of predominance of the Fe^{3+} ions shrinks when the pH decreases.

NOTE.– In constructing Figure 4.4, we have not taken account of the interaction between the Fe^{3+} ions and the water.

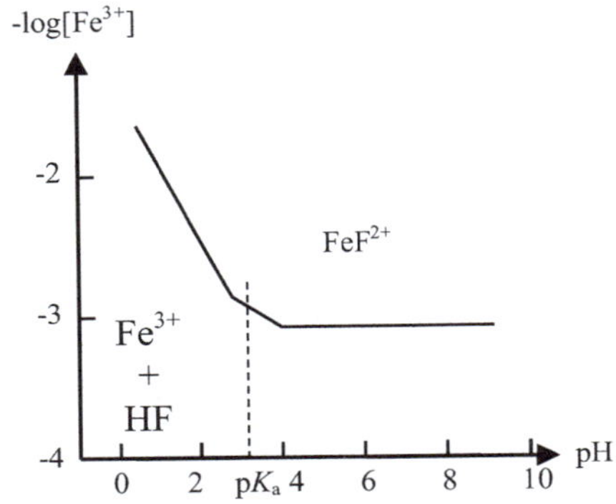

Figure 4.4. *Influence of the pH on the domains of predominance of the components of the FeF-complex*

4.1.2. Competition between two ligands on the same acceptor

Consider two complexes with the same central atom and two different ligands: ML and ML'. Their dissociations are written as:

$$ML = M + L \quad [4R.5]$$

$$ML' = M + L' \quad [4R.6]$$

The dissociation constants are expressed in the following forms:

$$K_{d1} = \frac{[M]\,[L]}{[ML]} \quad [4.12]$$

and

$$K_{d2} = \frac{[M]\,[L']}{[ML']} \qquad [4.13]$$

To find which ligand has the greater tendency to attach to the central atom, we consider the equilibrium:

$$ML' + L = ML + L' \qquad [4R.7]$$

Its equilibrium constant will be:

$$K_6 = \frac{[ML]\,[L']}{[ML']\,[L]} = \frac{K_{d2}}{K_{d1}} \qquad [4.14]$$

Thus, if $K_6 > 1$, the complex ML will be favored. This will happen if $K_{d2} > K_{d1}$.

On the other hand, if $K_6 < 1$, the complex ML' will be favored. This will be seen if $K_{d2} < K_{d1}$.

Hence, the favored complex is whichever one has the lower dissociation constant.

The same method can be employed to examine competition between two acceptors for the same ligand. The result would be identical.

Later on, we shall see the influence of the formation of complexes on the solubility of a salt (see section 5.2) and on the redox potential (see section 7.7.4).

The calculations of the concentrations in solutions involving complexes are performed in the same way as for acid–base reactions. In the approximate manual calculation method, we first look for a thermodynamically-equivalent system if the species have no domain of overlap; then we look for the equilibrium with the highest progress rate, which will be the control equilibrium.

4.1.3. *Method for studying perfect complexes*

For a perfect complex, only the coordination number n is of interest when trying to characterize the complex. Thus, study methods will try to determine that number; however some methods yield a value of the dissociation constant which is generally very low.

We distinguish between three families of methods:

– thermochemical methods;

– electrochemical methods;

– absorption spectrometry.

4.1.3.1. *Thermochemical methods*

The main thermochemical method is based on the examination of the solubility. We shall describe this method in section 5.3 of Chapter 5.

4.1.3.2. *Electrochemical methods*

Electrochemical methods are based on potentiometry (measurement of potential). We shall discuss these methods in section 7.7.5 of Chapter 7, as an application of galvanic cells.

4.1.3.3. *Absorption spectrometry*

The spectroscopic method is by far the most widely used method to characterize a perfect complex.

Consider a complex liable to decompose according to the following reaction:

$$ML_n = M + nL \qquad [4R.8]$$

The complex ML_n, which has a different structure to the components M and L, often exhibits properties of light absorption in the ultra-violet or visible spectrum (complexes are often colored) at wavelengths where the individual components exhibit no (or very little) absorption.

This light absorption obeys the Beer–Lambert law, whereby the optical depth is proportional to the wavelength of the solution through which the light is passed, and the concentration of that solution. The proportionality coefficient is the *coefficient of absorption* at the wavelength in question, ε_λ. Thus, the law is expressed as follows:

$$\log\frac{I_0}{I} = d = \varepsilon_\lambda lC \qquad [4.15]$$

We work with a constant concentration of M and add ever-increasing concentrations x of L.

We measure the optical depth, and by plotting the curve showing optical depth as a function of x, we obtain the curve in Figure 4.5. This figure is plotted for a scenario where the complex is more absorbent than its components at the chosen wavelength.

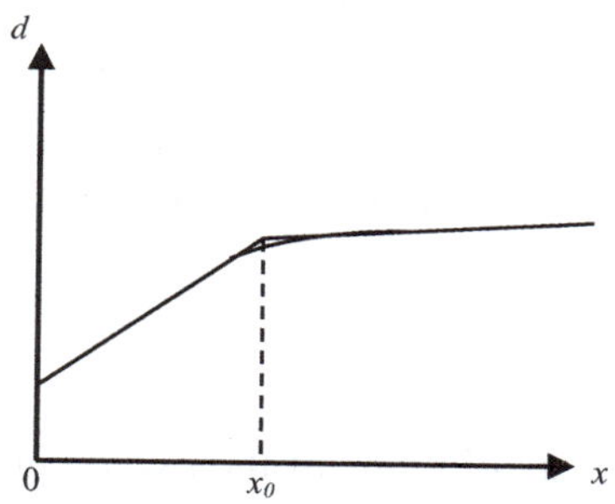

Figure 4.5. *Finding the equivalence by measuring the optical depth of absorption*

The abscissa of the inflection point x_0 is obtained by the intersection of the extension of both branches. That value x_0 gives us the number of moles n of L bound to a mole of M.

NOTE.– When multiple complexes are present, we may obtain several successive breaks. However, in such a case, the method is not very accurate. In section 4.4.5, we shall see a method which lends itself better to the study of successive complexes.

It sometimes happens that various types of ligands exist in the complex. In that case, the dissociation is written:

$$ML_nL'_{n'} = M + nL + n'L' \qquad [4R.9]$$

We shall make the decision to keep the concentrations of the two components M and the ligand constant whilst working, so as to maintain compatibility with the previous case. The study will be supplemented by measurement, keeping the concentration of the other ligand constant as well as that of M.

4.1.4. *Methods for studying imperfect complexes*

An imperfect complex shall be completely defined by the two values K_d and n. The above methods cannot be used to determine those two values. Two methods tend to be used: the *method of continuous variation* and the *corresponding solution method.*

4.1.4.1. *Method of continuous variation*

This method, which was developed by Job [JOB 28] is based on a general property of a chemical equilibrium. It states that the yield of the reaction is greatest at equilibrium where the reactants are mixed in stoichiometric proportions.

We operate as follows:

– we mix equimolecular solutions of the two components M and L which combine in accordance with the reaction [4R.8] to form the complex ML_n;

– let C denote the concentration common to these two solutions, and p the proportion of solution of L used in the mixture.

We can write the law of mass action for the equilibrium of formation of the complex:

$$K_d = \frac{[M][L]}{[ML_n]} \tag{4.16}$$

If $[M]_0$ and $[L]_0$ denote the total initial quantities of M and N, the conservations of the elements M and L enable us to write, respectively:

$$[M]_0 = [M] + [ML_n] = (1\text{-p})C \tag{4.17a}$$

$$[L]_0 = [L] + n[ML_n] = pC \tag{4.17b}$$

We know that the concentration of ML_n will be maximal at equilibrium when the two components M and L are mixed in the ratio 1:*n*, equal to the stoichiometric ratio, when *p* takes the value p_m such that:

$$\frac{p_m}{(1\text{-}p_m)} = n \qquad [4.18]$$

To experimentally determine the value of p_m, we generally use light absorption at an appropriate wavelength which is optimal for the complex. According to the Beer–Lambert law, the mixture's optical depth of the complex of its components is:

$$\frac{d}{lC} = \varepsilon_{\mathrm{ML}_n}\left[\mathrm{ML}_n\right] + \varepsilon_M(1\text{-}p) + \varepsilon_L p \qquad [4.19]$$

If the complex were not formed, the optical depth would be such that:

$$\frac{d_{th}}{lC} = \varepsilon_M(1\text{-}p) + \varepsilon_L p \qquad [4.20]$$

If we measure the difference Δ between the two values, then in light of relations [4.17a] and [4.17b], we obtain:

$$\Delta = \left(\frac{d}{lC} - \frac{d_{th}}{lC}\right) = \varepsilon_{\mathrm{ML}_n} - \left(\mathrm{n}\varepsilon_{\mathrm{M}} + \varepsilon_L p\right) \qquad [4.21]$$

For a given concentration *C*, we see that the maximum of the concentration of X corresponds to the peak of the curve showing the difference Δ.

Figure 4.6, as a function of the ratio *p*, shows the shape of the experimental curve measured by absorption and the calculated curve that we would obtain if there were no complexation reaction. By finding the difference on a point-by-point basis, we deduce the curve giving Δ as a function of *p*.

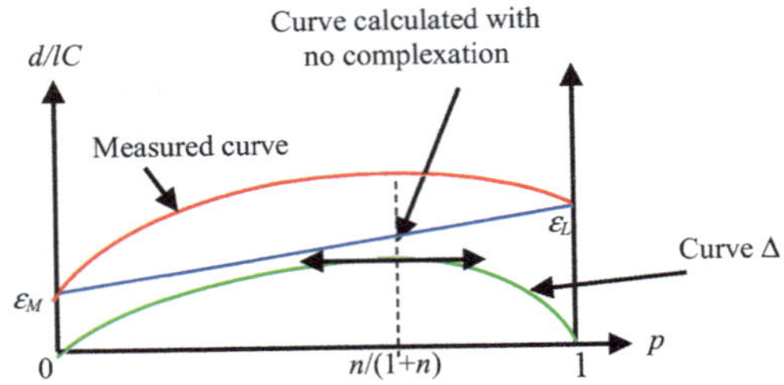

Figure 4.6. *Optical densities with the method of continuous variations method. For a color version of the figure, see www.iste.co.uk/soustelle/equilibria.zip*

The maximum of the curve Δ then gives p_m, and thus n, according to relation [4.18].

We now need to determine the second characteristic value of our complex: its dissociation constant K_D.

We proceed in the same way as above by mixing solutions of M and L, but this time we choose non-equimolecular solutions. The first will contain a concentration C of M and the second a concentration βC of L, such that β is very different to 1. We mix the two solutions in the ratio p. As in the previous case, there is a value of p which corresponds to a maximum of the concentration of ML_n, but that value is no longer p_m but instead a different value, written as p_β.

We always have the law of mass action given by the relation [4.16].

The conservations of the elements M and L are now written:

$$[\mathrm{M}]_0 = [\mathrm{M}] + [\mathrm{ML}_n] = (1\text{-}p)C \quad [4.21a]$$

$$[\mathrm{L}]_0 = [\mathrm{L}] + n[\mathrm{ML}_n] = p\beta C \quad [4.21b]$$

By differentiating in relation to p and combining relations [4.16], [4.21a] and [4.21b], in the knowledge that d[ML_n]/dp is 0, we obtain:

$$n\beta[\mathrm{M}] - [\mathrm{L}] \quad [4.22]$$

By replacing the values of the concentrations of M and L derived from equations [4.21a] and [4.21b], we deduce:

$$[\mathrm{X}] = \frac{\beta C}{1-\beta}\left[p_\beta\left((1+\frac{1}{n})-1\right)\right] \quad [4.23]$$

Measuring p_β by the same spectroscopic method to calculate the concentration of ML_n. The application of relations [4.21a] and [4.21b] then respectively allows us to calculate the concentrations of M and L. Those values are given in relation [4.16] to calculate the dissociation constant.

NOTE.– During the second phase, we took the precaution of mixing non-equimolecular solutions, in a proportion of concentrations β different to 1. Indeed, if we choose $\beta = 1$, because of relation [4.18], we see that relation [4.23] yields an indeterminate concentration of ML_n.

Note that if several complexes are liable to form, the value of the maximum p_β depends on the absorption coefficients, and thus on the wavelength. The method is no longer usable to determine the values n and K_d. However, that remark may reveal the existence of multiple complexes, which differ from one another by the number of ligands attached to their central atom.

4.1.4.2. *Corresponding solution method*

We shall prepare two series of solutions. Each series will be characterized by the concentration of M introduced, denoted by C. Each solution will be characterized by the concentration x of L introduced.

For each series of solutions, the balance of M is written:

$$[\mathrm{M}] + [\mathrm{ML}_n] = C \quad [4.24]$$

For each solution, the balance of the ligands is written:

$$[\mathrm{L}] + [\mathrm{ML}_n] = x \quad [4.25]$$

Each solution obviously obeys relation [4.16].

By choosing an acceptable wavelength – if possible, absorbing only the complex – we measure the optical depth of each solution. Based on the Beer–Lambert law, we can write the following for each solution i belonging to the series j:

$$\frac{d_{i,j}}{C_j l}=\frac{\varepsilon[\mathrm{ML}_n]}{C_j} \qquad [4.26]$$

For each of the two series j = 1 or 2, we plot the ratio y = d/Cl as a function of x (Figure 4.7).

We plot a horizontal for any given value of the ordinate y_0. That horizontal cuts each of the two curves at the points x_1 for the series of solutions 1 and x_2 for the series of solutions 4.

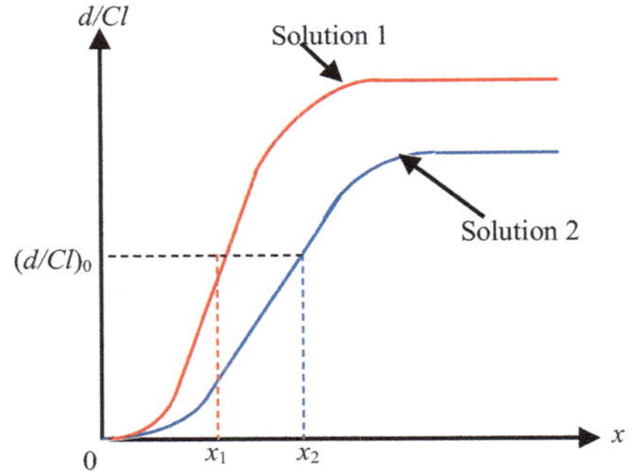

Figure 4.7. *Absorption by two corresponding solutions. For a color version of the figure, see www.iste.co.uk/soustelle/equilibria.zip*

By virtue of relation [4.26], at those points, the two solutions have the same ratio $[\mathrm{ML}_n]/C$, and thus by virtue of relation [4.24], the same concentration [M]. We say that these two solutions are *corresponding solutions*. According to relation [4.16], the two corresponding solutions have the same concentration of ligand [L].

Thus, we can write:

$$\frac{[\mathrm{ML}_n]_1}{C_1}=\frac{[\mathrm{ML}_n]_2}{C_2}=\frac{x_1-[\mathrm{L}]}{C_1}=\frac{x_2-[\mathrm{L}]}{C_2}=\frac{x_1-x_2}{n(C_1-C_2)} \qquad [4.27]$$

From this, we deduce:

$$[\mathrm{L}] = x_1 - C_1 \frac{x_1 - x_2}{(C_1 - C_2)} \qquad [4.28]$$

Using relation [4.26], we calculate the concentration of complex, and then from the second equation [4.27], we deduce *n*. Based on relation [4.24], applied to solution 1, for example, we deduce the concentration of M. Finally, based on relation [4.16], we calculate the dissociation constant.

4.1.5. *Study of successive complexes*

Up to this point, we have supposed that the association of the central atom M and the ligand L gave rise to only one complex. In the case that various successive combinations are possible, the above method can only be applied if the absorption wavelengths of the different complexes formed are sufficiently different to one another to allow us to isolate the absorption of each composition. Then, the experiments described above can be repeated, and enable us to determine the values of the different coordination numbers and of the overall dissociation constants corresponding to each composition. From these overall constants, we deduce the constants pertaining to each addition reaction.

If it is not possible to distinguish between the different compositions using a spectrograph, we then need to turn to a different method.

Thus, we consider that a central atom M, with ligands L, gives a succession of complexes with increasing coordination numbers, according to the equilibria:

$$\begin{aligned} &\mathrm{M} + \mathrm{L} = \mathrm{ML} \\ &\mathrm{M} + \mathrm{ML} = \mathrm{ML_2} \\ &\mathrm{M} + \mathrm{ML_2} = \mathrm{ML_3} \\ &\dots\dots\dots\dots \end{aligned} \qquad [4R.10]$$

These equilibria give us the following expressions of the law of mass action:

$$\begin{aligned}&[\mathrm{ML}]=K_{a1}[\mathrm{M}]\,[\mathrm{L}]\\&[\mathrm{ML}_2]=K_{a2}[\mathrm{M}]\,[\mathrm{ML}]\\&[\mathrm{ML}_3]=K_{a3}[\mathrm{M}]\,[\mathrm{ML}_2]\end{aligned}\qquad[4.29]$$

If the elements M are introduced at a concentration of C moles per liter, the conservation of elements M means that we can write:

$$[\mathrm{M}]+[\mathrm{ML}]+[\mathrm{ML}_2]+[\mathrm{ML}_3]+\ldots=C\qquad[4.30]$$

If the elements L are introduced at a concentration of x moles per liter, the same law for L gives us:

$$[\mathrm{L}]+[\mathrm{ML}]+2[\mathrm{ML}_2]+3[\mathrm{ML}_3]+\ldots=x\qquad[4.31]$$

As we did in section 4.1.4.2, we shall consider two corresponding solutions (1) and (2) – i.e. which have the same ratio d/C. Therefore, they contain the same proportion of M in a complex and the same [M]/C ratio.

Relation [4.30] shows that for these two corresponding solutions, we must have:

$$\frac{[\mathrm{ML}]_1}{C_1}+\frac{[\mathrm{ML}_2]_1}{C_1}+\frac{[\mathrm{ML}_3]_1}{C_1}+\ldots=\frac{[\mathrm{ML}]_2}{C_2}+\frac{[\mathrm{ML}_2]_2}{C_2}+\frac{[\mathrm{ML}_3]_2}{C_2}+\ldots\qquad[4.32]$$

So that this equality is preserved whatever the values of C and x, it is necessary that all the terms are identical, pair by pair, so:

$$\frac{[\mathrm{ML}]_1}{C_1}=\frac{[\mathrm{ML}]_2}{C_2};\ \frac{[\mathrm{ML}_2]_1}{C_1}=\frac{[\mathrm{ML}_2]_2}{C_2};\ \frac{[\mathrm{ML}_3]_1}{C_1}=\frac{[\mathrm{ML}_3]_2}{C_2};\ldots\qquad[4.33]$$

We can define the mean coordination number, which is applicable to each of the two solutions, by the relation:

$$\bar{n}=\frac{[\mathrm{ML}]}{C}+2\frac{[\mathrm{ML}_2]}{C}+3\frac{[\mathrm{ML}_3]}{C}+\ldots\qquad[4.34]$$

From equations [4.29], [4.33] and [4.34], we deduce that the concentration of free ligand [L] and the mean coordination number are identical for the two solutions, and that we can write:

$$\bar{n} = \frac{x_1 - [L]}{C_1} = \frac{x_2 - [L]}{C_2} = \frac{x_1 - x_2}{C_1 - C_2} \qquad [4.35]$$

From these relations, and based on the knowledge of the values of x_1, x_2, C_1 and C_2, we can easily deduce the values of $\bar{n}$ and $[L]$.

4.2. Redox reactions

Oxidation reactions occur between ions – usually in water. In order to define redox reactions, we need to be familiar with the concept of the degree of oxidation – a value which is calculated on the basis of an electronegativity scale.

4.2.1. *Electronegativity – electronegativity scale*

The idea of electronegativity of elements is of prime importance in understanding the ionic nature of the bonds and thus laying the foundations for understanding reactivity. It is also important for the definition of what we call the *degree of oxidation*, or *oxidation number*, of an element in a compound.

4.2.1.1. *Qualitative definition of an electronegativity scale*

Looking at the bond between the H and F atoms in hydrogen fluoride, we see that it is formed by a pair of electrons (a doublet), one of which comes from the 1s orbital of the hydrogen atom and the other from the 2s orbital of the fluorine. We know that this bond exhibits an electrical moment, which means that the doublet is captured completely by the fluorine, thus leading to the existence of two ions: F^- and H^+. We say that fluorine is more electronegative than hydrogen. In other words, we say that an element A is more electronegative than an element B if A attracts the doublet of electrons, forming the A-B bond to itself.

Now let us consider an element which is far removed from fluorine in the periodic table – cesium, for example. When reacted with hydrogen, it forms a hydride with the formula CsH. Note that this bond also displays a strong electrical moment, but that this time, it is the hydrogen atom which captures the electron doublet forming the bond, thereby leading to the formation of the ions Cs^+ and H^-. We say that hydrogen is more electronegative than cesium.

By comparing the two compounds FH and CsH, we can deduce that fluorine is more electronegative than cesium.

Now consider the bond between carbon and hydrogen in methane. This bond exhibits only a slight electrical moment. The carbon attracts the bond pair slightly towards itself, but it does not go so far as to form ions. Thus, carbon is slightly more electronegative than hydrogen. Hence, we can begin to construct a scale showing the elements in order of electronegativity: Cs, H, C, F, in order of increasing electronegativity.

Not all elements actually form a bond with hydrogen that would allow us to classify their electronegativity. However, it is possible to examine their bonds with other elements, and if the latter elements form bonds with hydrogen atoms, we can still manage to construct our scale, step by step. It is quite common to use bonds with oxygen as a comparison point, as we know – as is demonstrated by the bonds in the water molecule – that oxygen is significantly more electronegative than hydrogen, but we cannot expect so great an electronegativity value as found in fluorine.

For the time being, this scale remains qualitative. Very early on, scientists came up against the need to quantify the idea. Therefore, a number of quantitative scales were put forward. The three most significant such models were developed by Pauling, Mulliken, and Allred and Rochow.

4.2.1.2. *Pauling's electronegativity scale (1932)*

In 1932, Pauling put forward the idea of a scale based not on the electrical moment of the bonds, as suggested by the above definition of electronegativity, but instead on the bond energies. Yet it should be noted that *a priori* the bond energies have no obvious connection with the electrical moments.

Pauling supposed that for a bond without an electrical moment – i.e. one which is completely symmetrical between the atoms X and Y – the energy contained in the XY bond was the arithmetic mean of the bond energies in the associations X_2 and Y_2, so:

$$D_{XY} - \frac{1}{2}(D_{XX} + D_{YY}) = 0 \qquad [4.36]$$

In the case of a non-symmetrical bond, where an electrical moment is present, Pauling looked at the difference Δ defined by:

$$\Delta = D_{XY} - \left(\frac{D_{XX} + D_{YY}}{2}\right) \qquad [4.37]$$

He noted that, for instance, with HF, we have Δ = 15.31 kJ/mole, whilst for IH, we have 0.38 kJ/mole. However, what is certain is that iodine is less electronegative than fluorine. Thus, it seems that we can envisage a relation between the value Δ and the difference between the electronegativity of elements X and Y: $el_X - el_Y$. After several rounds of trial and error, Pauling, for the value of Δ expressed in eV/atom, chose the relation:

$$el_X - el_Y = \sqrt{\Delta} \qquad [4.38]$$

Often, though, bond energies are expressed and quoted in kJ/mole, so we need a conversion factor which transforms relation [4.38] into:

$$el_{X(P)} - el_{Y(P)} = 0.102\sqrt{\Delta} \qquad [4.39]$$

Choosing an arbitrary value for an element, Pauling chose $el_{H(P)} = 0$. Thus, he established his first quantitative scale. Unfortunately, on this scale, practically all metals are found to have negative electronegativity values. To obtain a positive value for all elements, noting that the strongest electronegativity is that of fluorine, he assigned fluorine the value $el_{F(P)} = 4$. Thus, the electronegativity of hydrogen becomes $el_{H(P)} = 2.1$. Pauling added 2.1 to all the values extracted by the application of relation [4.39]. We deduce that the electronegativity of an element X, on Pauling's scale, is given by:

$$el_{X(P)} = 2.1 + 0.102\sqrt{\Delta_{XF}} \qquad [4.40]$$

Δ_{XF} is the difference [4.38] for the bond between fluorine and the atom X under study.

Thus, a complete quantitative scale is constructed. Table 4.1 shows an extract from that scale. As we can see, according to Pauling, practically all the transition elements have an electronegativity of 1.6.

NOTE.– To begin with, Pauling chose the geometric mean of the bond energies to define Δ, instead of the arithmetic mean used in relation [4.37]. The results have proven to be more reliable with the latter choice.

H	2.1	F	4.0	Cl	3.0	Br	2.8	I	2.4
Li	1.0	Na	0.9	K	0.8	Rb	0.8	Cs	0.7
Be	1.5	Mg	1.2	Ca	1.0	Sr	1.0	Ba	0.9
C	2.5	If	1.8	Ge	1.7	Sn	1.7		
				Ti	1.6	Zr	1.6		
N	3.0	P	2.1	As	2.0	Sb	1.8		
0	3.5	S	2.5	Se	2.4	Te	2.1		

Table 4.1. *Electronegativity table according to Pauling*

4.2.1.3. *Mulliken's electronegativity scale (1934)*

In 1934, Mulliken defined an electronegativity scale that fits more closely with the relations between an atom and an electron. These relations lie firstly in the electron affinity – i.e. the energy involved in the fixation of an electron by a neutral atom I – and secondly in the ionization energy – the energy released when an atom loses an electron ε_{ea}. Mulliken defined electronegativity as the arithmetic mean of the electron affinity and the ionization energy of the atom. As that mean may be positive or negative depending on the element, he chose the absolute value of that mean so as to obtain only positive electronegativity values. Thus, we have the following expression for Mulliken's definition of electronegativity:

$$el_{\mathrm{X(M)}} = \frac{\left|I + \varepsilon_{ea}\right|}{2} \qquad [4.41]$$

Note that the order in which elements are classified on Pauling's and Mulliken's scales is the same. Thus, we can switch from one scale to the other by setting, say, the value 4 for fluorine – a value which is equal to 252.4 kJ/mole on Mulliken's scale. The conversion from one scale to the other is achieved by a linear transformation, which, if the energies are expressed in eV/atom, is:

$$el_{\mathrm{X(M)}} = 0.187\left(I + \varepsilon_{ea}\right) + 0.17 \qquad [4.42]$$

If the energy values are expressed in kJ/mole, that transformation becomes:

$$el_{\mathrm{X(M)}} = 1.97 \times 10^{-3}\left(I + \varepsilon_{ea}\right) + 0.19 \qquad [4.43]$$

It is worth pointing out that the value defined by Mulliken is a value specific to the atom, whereas Pauling's is based on the property of the bond of the atom in a molecule.

Obviously, we can calculate the Mulliken electronegativity of an element only if we are able to find its electron affinity. Unfortunately, that value is not yet known for all elements in the periodic table. There are around thirty stable elements whose electron affinity is unknown.

4.2.1.4. *Allred–Rochow electronegativity scale (1958)*

Allred and Rochow defined the electronegativity of an element X as the electrical force exerted between the electron and a charge characterizing the rest of the atom. For that charge, they chose the effective charge Z_{eff} calculated using Slater's method (which will be discussed later on). The distance chosen for that force is the covalent radius of the element – i.e. half the distance between the atoms in the X-X molecule. Allred and Rochow's scale, therefore, is defined by the relation:

$$el_{\mathrm{X(AR)}} = F_X = \frac{Z_{eff\,\mathrm{(X)}}\,e^2}{r^2_{\mathrm{(X)}}} \qquad [4.44]$$

The scale obtained is, once again, the same as that given by the previous two definitions. We can switch from Pauling's scale to Allred and Rochow's (adopting the value 2.1 for hydrogen) by the following relation (r expressed in nm):

$$el_{\mathrm{X(AR)}} = \frac{35.9\,Z_{eff\,\mathrm{(X)}}}{r^2_{\mathrm{(X)}}} + 0.64 \qquad [4.45]$$

Admittedly, this last definition of electronegativity comes closest to the initial definition, which involves the direction of movement of the bond electrons. Regrettably, though, no scale has been constructed on the basis of the true definition – i.e. on the electrical moment of the bonds.

For an atom with the atomic number Z, the effective charge is calculated by the relation:

$$Z_{eff} = Z - \sigma \qquad [4.46]$$

Slater then gives us the means to calculate the shielding effect σ, as follows:

– for an electron situated in a 1s orbital, the shielding effect caused by the other 1s electron is equal to 0.30;

– for an electron situated in an ns or np orbital, where n is greater than 1, the shielding effect experienced because of each of the other electrons in that group is equal to 0.35;

– the shielding effect experienced by that same electron because of the electrons of the next level down is equal to 0.85 per electron;

– the shielding effect experienced by that electron because of the levels below the next level down is equal to 4.00 per electron;

– finally, electrons situated at higher energy levels than the one in question here have no shielding effect whatsoever on the electron under consideration.

4.2.1.5. *Electronegativity and the periodic table*

If we examine the evolution of the electronegativity of the elements as listed in the periodic table, we note that, as Figure 4.8 illustrates:

– in any given column, electronegativity increases as we move upwards, because the atomic number decreases. The atomic nucleus has a tendency to attract valence electrons which are less well shielded by the core electrons;

– in given row, electronegativity increases as we move to the right, because the increase in the atomic number means an increase in the charge carried in the nucleus (a greater number of protons), which thus interacts more strongly with the valence electrons.

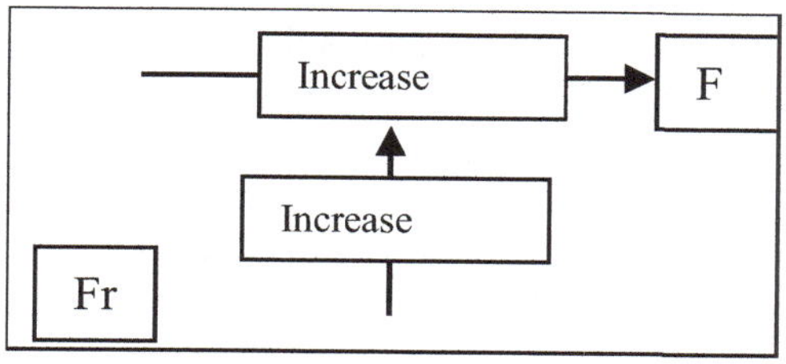

Figure 4.8. *Changing values of electronegativity in the periodic table of the elements*

The minimum electronegativity, therefore, is at the bottom left of the table, with francium. The maximum is at the top right, with fluorine.

Figure 4.9 shows the electronegativity values, on Pauling's scale, as a function of the place on the periodic table. This figure confirms the observations we have just made. In addition, it demonstrates that the electronegativity of the first element in each column is similar to that of the elements in the column immediately to its right. Thus, the electronegativity of lithium is close to that of calcium. The same is true of beryllium and aluminum, and of boron and silicon, nitrogen and sulfur, oxygen and chlorine. Indeed, we can see great chemical similarity between the elements mentioned.

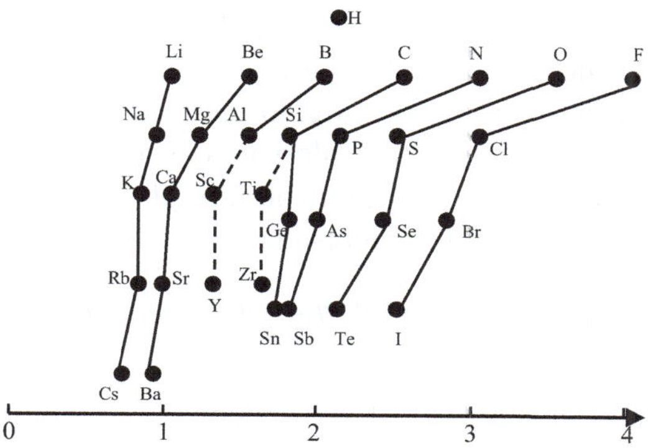

Figure 4.9. *Electronegativity and the periodic table*

Also note that in each column, the last element is on a par with the elements in the column situated immediately to its left. Thus, we note the proximity of the values of iodine and sulfur, tellurium and phosphorus, barium and rubidium. Similarly to above, we observe similarities in the chemical properties between the elements we compare.

NOTE.– Given that rare gases practically never form a compound, the notion of electronegativity is meaningless when dealing with them.

4.2.2. Degrees of oxidation

The idea of the degree of oxidation is crucially important when studying redox reactions and for all electrochemical thermodynamics. We shall approach it by way of binary ionic compounds, and then extend it to apply generally to all molecules.

4.2.2.1. Degrees of oxidation of the elements of a binary ionic compound

Consider solid potassium chloride. It is made up of chlorine ions Cl^- and potassium ions K^+. This separation into ions stems from the fact that chlorine is much more strongly electronegative than potassium, and therefore captures the two bond electrons, one of which comes from the potassium atom. We say that the degree of oxidation of the chlorine is -1 and that of the potassium is +1.

More generally, for ionic compounds, the degree of oxidation of an element is equal to the charge of the ion formed. For instance, in magnesium chloride, $MgCl_2$, which is made up of chlorine ions Cl^- and magnesium ions Mg^{2+}, the degree of oxidation of the chlorine is -1 and that of the magnesium is +2. Because of the rule of electrical neutrality of molecules, the sum of the degrees of oxidation of the two elements in an ionic binary compound is zero.

4.2.2.2. Degrees of oxidation of the elements in a covalent binary compound

Let us now look at the case of a covalent binary compound. We shall choose hydrogen chloride as a case study. Chlorine has greater electronegativity than hydrogen. It follows that the bond doublet between the two atoms is shifted to the side of the chlorine atom. By convention, we attribute both electrons from the doublet to chlorine, and conclude that in the HCl molecule, the degree of oxidation of the chlorine is -1 and that of the hydrogen is +1.

More generally, in a covalent binary compound, we conventionally attribute the two electrons in the doublet of a bond between the two atoms to the more strongly electronegative atom, and the degree of oxidation is given by the number of doublets attributed to (negative degree of oxidation) or ceded by (positive degree of oxidation) the atom in place. Of course, the sum of the degrees of oxidation in the molecule is always zero.

If the two atoms making up the binary compound are identical – as is the case in the O_2 molecule, for instance – the center of gravity of the charges is precisely the midpoint of the distance between the two nuclei (here oxygen). It is not possible to attribute doublets any more to one of the atoms than to the other. We say that the degree of oxidation of the oxygen in the O_2 molecule is zero. The same is true for all elements belonging to homonuclear binary compounds.

4.2.2.3. *Generalization of the concept of the degree of oxidation*

Now consider a more complex molecule. We shall choose the example of iron(III) chloride. The semi-structural formula of this compound is:

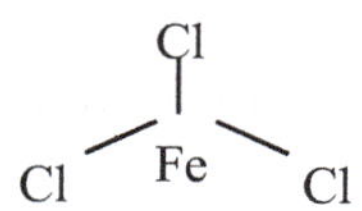

Examine each of the three bonds which, in this case, are identical. For a bond, as chlorine is more electronegative than iron, we attribute the corresponding doublet to the chlorine atom. Repeating the operation for each bond, we can say that in $FeCl_3$, the degree of oxidation of each chlorine atom is -1, whilst that of the iron is +3.

It may be that the same element has two different degrees of oxidation within the same molecule. Take the example of potassium persulfate, which has the formula $K_2S_2O_8$, and examine the degree of oxidation of the atoms making up that compound. In order to do so, we write its semi-structural formula:

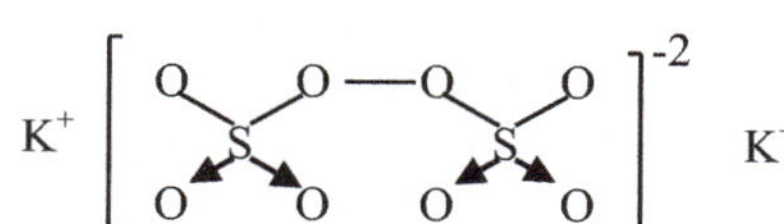

Three types of bonds come into play in this compound:

– firstly, the ionic bond between potassium and the persulfate anion, which enables us to attribute the degree of oxidation +1 to the potassium;

– four covalent bonds linking a sulfur atom and an oxygen atom. In each of them, as oxygen is more electronegative than sulfur, it receives both of the electrons in the doublet. The outer two atoms also receive the two electrons from the charge of the ion; thus, their degree of oxidation is -2 (one electron due to the charge of the ion, and the other to the bond with the sulfur atom). The two inner oxygen atoms are connected to one another by a symmetrical bond whereby the electrons of the doublet cannot be attributed to one atom or the other. Hence, these oxygen atoms have the degree of oxidation -1 (stemming from their bond with the sulfur);

– four (dative) coordination bonds, in which the two electrons in the doublet are donated by the oxygen atom, giving it the degree of oxidation -2. It follows from this that the sulfur atoms have the degree of oxidation +6.

The sum of the degrees of oxidation in the molecule is indeed 0, as the sum of the degrees of oxidation in the anion is -2.

4.2.2.4. *Practical calculation of the degree of oxidation of an element in a molecule*

In practice, we do not always use the semi-structural formula. We have at our disposal a certain number of known facts, which can be used to quickly calculate the degree of oxidation of an atom in a molecule on the basis of its simple formula:

– we can practically always attribute a degree of oxidation of +1 to all alkali metals (first column in the periodic table) and the degree +2 to all alkaline earth metals (second column of the periodic table);

– we can practically attribute the degree +1 to a hydrogen atom, except in alkali hydrides and alkaline earth hydrides, where the degree of oxidation of the hydrogen is -1;

– we can attribute the degree of oxidation -2 to an oxygen atom (with certain exceptions due to the presence of a peroxide bridge in persulfates and oxygenated water H_2O_2, which then have two oxygen atoms, each with the degree of oxidation -1):

– finally, we use the rule of nullity of the sum of the degrees of oxidation of a molecule.

We choose the example of potassium permanganate, whose overall formula is $KMnO_4$. We attribute the degree of oxidation +1 to the potassium (alkali metal), the degree -2 to the oxygen atoms, and as the sum of the degrees of oxidation must be zero, we are led to attribute the degree of oxidation +7 to the manganese.

Of course, the rules of calculating using the simple formula can sometimes lead to incorrect results. Look again at the example of potassium persulfate and work purely on the basis of the simple formula, remaining unaware of the existence of the peroxide bridge. In applying the above rules, we immediately find the degree +1 for the potassium, and -2 for the oxygen atoms. Now, when we apply the balance law, for each of the two sulfurs, we find the degree of oxidation +7, which is obviously wrong.

Sometimes, the result obtained by using the basic chemical formula may be not only false, but strange as well. Take the example of sodium tetrathionate $Na_2S_4O_8$. The attribution of the degree of oxidation +1 to sodium and -2 to oxygen immediately gives us the value 2.5 for the degree of oxidation of each sulfur atom. This result is obviously ridiculous – it is impossible to attribute half an electron to an atom! The result thus obtained is, in fact, the mean between the true degrees of oxidation of the four sulfur atoms: +6, +6, -2 and 0. This is what we call the *crude degree of oxidation.*

The fractional result of a degree of oxidation of an element in a compound is always the result of a mean which expresses the existence of different degrees of oxidation for atoms of the same element.

4.2.2.5. *Degree of oxidation of a coordinating element in a complex ion*

Complex ions present a difficulty when we want to determine the degree of oxidation of the coordinating element.

Determining the degree of oxidation of the coordinating element in a complex may lead to a number of difficulties which, fortunately, can be resolved by a convention. Remember, first of all, that the formula of a complex ion can be written in two different ways: in the form of the nucleus bound to each of the ligands, or in the form of an additive compound. If we take the example of hexammine cobalt(III) chloride, the two written forms would be $[Co[NH_3]_6]^{3+}$ or $CoCl_3,6NH_4$. If we take the structural formula, we would quickly see that the CO^{3+} ion, through a dative bond, receives $2 \times 6 = 12$ electrons from the ammonia. However, as it had already donated three electrons to that ammonia, we can conclude that the degree of oxidation of the cobalt in that compound is $12 - 3 = 9$, which is formally accurate. Yet in reality, it is very difficult to distinguish dative bonds from purely electrostatic bonds, where the degree of oxidation of cobalt remains unchanged, and is therefore 2. To circumvent this problem, we say that the degree of oxidation of the central ion, in a complex, is the degree which that element has in the corresponding simple compounds – i.e. the degree of oxidation that would be attributed to that element if writing the complex in the form of an additive compound. In this case, for our example, we immediately find a degree of oxidation +3 for the cobalt, corresponding to the simple compound $CoCl_3$.

4.2.3. *Definition of redox reactions*

A redox reaction is a reaction in which two elements have their degrees of oxidation changed. In order to respect the electrical neutrality of the elements, the sum of the variations in the degrees of oxidation must be zero. One element sees its degree of oxidation increase: it is said to be *reduced* by the reaction. The other element has its degree of oxidation decrease, and is said to be *oxidized* by the reaction. The compound containing the element reduced by the reaction is the *oxidant* (or oxidizing agent); that which contains the element that is oxidized is known as the *reductant* (or reducing agent). Thus, the redox reaction looks like this:

$$\text{oxidant} + \text{reductant} \rightarrow \text{reduced form} + \text{oxidized form}$$

NOTE.– A redox reaction is often presented as being an electron-exchange reaction. In actual fact, though, very often – particularly with reactions involving covalent bonds – there is no real exchange of electrons.

4.2.4. *The two families of redox reactions*

When we examine the different redox reactions taking place in an aqueous medium, we can distinguish between two categories: reactions which do not involve the ions of water, and those which do.

The first family is represented by simple reactions between an oxidant and a reductant on their own. By way of example, we can cite the reduction of copper ions by iron, as follows:

$$Fe + Cu^{2+} \rightarrow Fe^{2+} + Cu \qquad [4R.11]$$

This reaction occurs spontaneously, when we simply place a few fragments of steel (e.g. a few pinches of shavings) in a copper sulfate solution.

The law of mass action for this equilibrium is written as:

$$K = \frac{|Fe^{2+}|^2}{|Cu|^2} \qquad [4.47]$$

The pH has no influence on the conditions at equilibrium.

The second family is more complex, as it involves water and its ions H^+ and OH^- , or rather H_3O^+ and OH^-, as the H^+ ion is unstable in water. For example, we can cite the reaction of reduction of potassium permanganate by tin(II) salts in an acid medium:

$$2MnO_4^- + 5Sn^{2+} + 16H_3O^+ \rightarrow 2Mn^{2+} + 5Sn^{4+} + 24H_2O \qquad [4R.12]$$

In this case, the law of mass action takes the form:

$$K = \frac{|Mn^{2+}|^2 |Sn^{4+}|^5}{|MnO_4^-|^2 |Sn^{2+}|^5 |H_3O^+|^{16}} \qquad [4.48]$$

The influence of the pH on the concentrations at equilibrium is obvious.

This second family is characterized by the presence of atoms of oxygen and/or hydrogen in one (or both) of the reagents.

It is common to represent a general redox equilibrium in the form of the reaction:

$$n_1 Ox_1 + n_2 Red_2 \rightarrow n'_2 Ox_2 + n'_1 Red_1 \qquad [4R.13]$$

We can see that whilst this formulation is perfectly valid for simple reactions, in the case of reactions involving the ions in water, it is necessary to include those ions on both sides of the equation, in order to apply the law of mass action. Symbolically, they are not mentioned, and we apply the law of mass action adopting the following formula regardless:

$$K = \frac{|\mathrm{Ox}_2|^{n'_2} |\mathrm{Red}_1|^{n'_1}}{|\mathrm{Ox}_1|^{n_1} |\mathrm{Red}_2|^{n_2}} \qquad [4.49]$$

4.2.5. *Dismutation and antidismutation*

Sometimes, the oxidant and the reductant reacting with it belong to the same molecule, in which one element shall have its oxidation number increase, and another element of the same nature in a different place in the molecule will see its oxidation number decrease. In such a case, we say that we have a *dismutation* of the element in question. For instance, oxygenated water is unstable because of its spontaneous decomposition into oxygen and water, which is a dismutation, according to the reaction:

$$2H_2O_2 \rightarrow O_2 + 2H_2O \qquad [4R.14]$$

Oxygenated water contains 2 oxygen atoms, each with the degree of oxidation -1, owing to the presence of an oxygen bridge. The degree of oxidation of one of those atoms from the bridge obtains the degree of oxidation 0 (oxygen gas), whilst the other obtains the degree -2 in the water molecule.

The reverse phenomenon – where the oxidant element and the reducing element, which initially belong to two different molecules, form a unique molecule containing an element with one single degree of oxidation – is known as *antidismutation* or *amphoterization*. We can cite the example of the oxidation of the iodide ion by the iodate ion, as follows:

$$IO_3^- + 5I^- + 6H_3O^+ \rightarrow 3I_2 + 9H_2O \qquad [4R.15]$$

In this reaction, iodine, which has the degree of oxidation +5 in the iodate ion and -1 in the iodide ion, ultimately obtains a degree of oxidation 0 in di-iodine.

4.2.6. *Redox reactions, and calculation of the stoichiometric numbers*

Note that for a simple redox reaction, whilst it is easy to find the stoichiometric numbers merely by balancing the charges, the same is absolutely not true with complex reactions. In complex reactions, as well, those stoichiometric numbers are not always elementary, as is shown by the example of reaction [4R.11]. We shall describe a systematic method to determine the stoichiometric numbers of a redox reaction involving water and its ions. That method is based on the differences between the degrees of oxidation of a given element between the two sides of the reaction.

To describe the method, we shall use the example of the reduction of permanganate by tin(II) ions. We write the reaction in the form:

+2

$$xMnO_4^- + ySn^{2+} + zH_3O^+ \rightarrow xMn^{2+} + ySn^{4+} + wH_2O$$

-5

First of all, we determine x and y. To do so, we calculate the variation of the oxidation number of the two elements concerned. In our example, those elements are manganese and tin.

The degree of oxidation of the manganese falls from +7 to +2, giving us a variation of the absolute value of its degree of oxidation:

$$|\Delta_{DOX}(Mn)| = 5$$

The tin's degree of oxidation rises from +2 to +4, which gives us a variation of the absolute value of its degree of oxidation:

$$|\Delta_{DOX}(Sn)| = 2$$

The lowest common multiple of these two variations is 10. The variations in each of the degrees of oxidation must be the same, and equal to that lowest common multiple. Thus, we adjust the values of x and y. In our example, we can deduce the values: $x = 2$ and $y = 5$.

We then determine the value of z. In order to do so, we balance out the electrical charges between the two sides of the equation, by adding

H_3O^+ ions on the appropriate side. In our case, we need to add these ions on the left-hand side, and we obtain $z = 16$.

Finally, we determine the value of w. In order to do so, we balance out the hydrogen atoms by placing water molecules on the correct side. In our example, we immediately obtain $w = 24$, and place those molecules on the right-hand side.

We can check our calculation by counting the number of oxygen atoms on each side of the equation. If our calculations are correct, we should obviously find the same number on both sides. In the present case, there are indeed 24 atoms of oxygen on each side of the equation.

The balancing method presented here applies when we content ourselves with crude, or even fractional, degrees of oxidation. This renders the idea of the crude degree of oxidation interesting, because we can write a reaction without needing to know the details of the internal bonds in the molecules in question.

4.2.7. Concept of a redox couple

In the same way as acid–base reactions appear to be the superposition of two half-reactions involving H^+ ions despite the instability of that ion in water, a redox reaction can be broken down into two half-reactions involving electrons, although they too are unstable in water. We obtain redox couples – for example, the redox couple Sn^{+4}/Sn^{+2} would give us the half-reaction:

$$Sn + 4 + 2e = Sn + 2 \qquad [4R.16]$$

We generalize this formulation, by convention, by placing the electrons on the left hand side, in the form:

$$\text{Oxidant} + ne \rightarrow \text{Reductant} \qquad [4R.17]$$

If the ions from water play a part, we write the equation for the couple and preserve the H^+ ions; for instance, for the manganese/permanganate ion couple, we write:

$$Mn^{2++} + 5e^- + 4H_2O \rightarrow MnO_4^- + 8H^+ \qquad [4R.18]$$

These reactions are, in fact, electronic reactions, which we shall study in Chapter 6, in the case that the electrons are stable because they are captured by an electronic conductor.

To balance a half-reaction relative to a redox couple, we first determine the number of electrons n involved, which is simply the absolute value of the difference between the degrees of oxidation of the oxidized form and the reduced form $|\Delta_{DOX}|$. Next, we balance the electrical charges using protons, if necessary, and finally balance out the number of oxygen atoms by adding water molecules.

From this concept of the redox couple, we deduce a second method for balancing redox reactions. We see that the formulation of the balance equation of a redox reaction such as reaction [4R.11] is the linear combination of the two reactions of redox couples, and possibly that of solvation of the proton:

$$H_2O + H^+ \rightarrow H_3O^+ \qquad [4R.19]$$

This linear combination is that which eliminates the electrons and protons. Thus, reaction [4R.12] is the combination of reactions [4R.16], [4R.18] and [4R.19] such that:

$$[4R.12] = 2[4R.18]-5[4R.16]-8[4R.19]$$

We can see that it is easy to determine the coefficients in the two equations for the redox couples (here 5 and 2) on the basis of the lowest common multiple of the absolute values of the degrees of oxidation. The value of the third coefficient (here 8) is determined immediately by elimination of the protons.

When two couples are present, it is interesting to find out which direction the reaction will spontaneously occur in. We shall answer this question in Chapter 6 (see section 6.5), once we have looked at redox potentials. Those potentials will also enable us to determine the standard Gibbs energy of a redox reaction (see section 7.7.9).

5

Precipitation Reactions and Equilibria

The majority of precipitation reactions are reactions between ions in solution which trigger the precipitation of a solid phase. Such reactions lead to states of equilibrium. For the most part, they are observed in water, but we may also see precipitations in other solvents, such as precipitations in molten-salt media, for instance.

5.1. Solubility of electrolytes in water – solubility product

In this section, we shall discuss an electrolyte composed of one anion and one cation, having the formula $A_{\nu_+}B_{\nu_-}$ for the non-dissociated molecule.

Suppose that this compound is not very soluble at equilibrium, reaching saturation at a low concentration. Thus we consider it to be totally ionized. We are in the presence of equilibrium [5R.1].

$$A_{\nu_+}B_{\nu_-}(solid) = \nu_+ A^{z_+ +} + \nu_- B^{z_- -} \qquad [5R.1]$$

As the solid phase is pure, the application of the law of mass action to that equilibrium yields the *solubility product* at infinite dilution K_s^∞:

$$\left|A^{z_+ +}\right|^{\nu_+} \left|B^{z_- -}\right|^{\nu_-} = K_s^\infty \qquad [5.1]$$

Taking account of the non-ideality of the solution, in the dilute media we are usually dealing with, the activities of the ions are linked to the concentrations by the expressions:

$$\left|A^{z_+ +}\right| = \gamma_+ \left[A^{z_+ +}\right] \qquad [5.2a]$$

and:

$$\left|B^{z_- -}\right| = \gamma_- \left[B^{z_- -}\right] \qquad [5.2b]$$

The product of the concentrations constitutes the solubility product K_S, which obeys the expression:

$$\left[A^{z_+ +}\right]^{\nu_+} \left[B^{z_- -}\right]^{\nu_-} = \frac{K_s^{\infty}}{\gamma_{\pm}^{(\nu_+ + \nu_-)}} = K_s \qquad [5.3]$$

We have assumed that the solution contains only one electrolyte. We shall see later on what happens to the solubility of that salt in the presence of other electrolytes.

5.2. Influence of complex formation on the solubility of a salt

Let us look at the $A_{\nu_+}B_{\nu_-}$ salt again. In the presence of pure water, the solubility of that salt is given by:

$$s = \left[A_{\nu_+}B_{\nu_-}\right]_{sat} = \left(K_s\right)^{\frac{1}{\nu_+ + \nu_-}} \qquad [5.4]$$

To the saturated solution, we add a complexing agent, which forms a complex with the cation $A^{z_+ +}$. We take this complex to be perfect and much more soluble than the original salt $A_{\nu_+}B_{\nu_-}$. The addition of the complexing agent shifts the equilibrium [5R.1] to the right, and thus increases the quantity of salt in solution, only this time, the cation is in complexed form. Thus, the quantity of the initial salt which it is possible to dissolve is determined by the solubility of the complex.

5.3. Application of the solubility product in determining the stability constant of complex ions

We saw in Chapter 4, section 4.1.3.1, that there is a thermochemical method based on the study of solubility to determine the dissociation constant and the coordination number of a complex ML_n. We shall now outline this method, which superimposes the equilibrium of dissociation of the complex on that of precipitation of a salt of the cation, with low solubility, involved in the complex.

Remember that the equilibrium of dissociation of the complex is written:

$$M^{z_+ +} + nL = ML_n^{z_+ +} \qquad [5R.2]$$

The application of the law of mass action to that equilibrium, working in the situation of a sufficiently dilute solution to be able to treat the activities and concentrations as one and the same, gives us:

$$\frac{\left[M^{z_+ +}\right][L]^n}{\left[ML_n^{z_+ +}\right]} = K_D \qquad [5.5]$$

If we add an anion $B^{z_- -}$ which is prone to form a salt with low solubility $M_{\nu_+}B_{\nu_-}$ with the cation $M^{z_+ +}$, that salt then precipitates. We can then add the ligand L. The salt $M_{\nu_+}B_{\nu_-}$ dissolves once again. In practical terms, when the salt is completely dissolved, the cation is entirely in the form of the complex, and if we overlook the solubility of the salt $M_{\nu_+}B_{\nu_-}$, we have:

$$\left[ML_n^{z_+ +}\right] \cong s \qquad [5.6]$$

because we have:

$$\left[M^{z_+ +}\right] << \left[ML_n^{z_+ +}\right] \qquad [5.7]$$

If we let x denote the excess concentration of ligand used, then we have:

$$[\mathrm{L}] = x \qquad [5.8]$$

By substituting relations [5.6] and [5.8] back into relation [5.5], we find:

$$\frac{\left[\mathrm{M}^{z_+ +}\right]x^n}{s} = K_D \qquad [5.9]$$

Using relation [5.3] in conditions of low concentration, we can write:

$$\left[\mathrm{M}^{z_+ +}\right] = \frac{K_s}{s} \qquad [5.10]$$

By superimposing the two equilibria [5R.1] and [5R.2], the two relations [5.3] and [5.9] give:

$$s^2 = \frac{K_D}{K_s} x^n \qquad [5.11]$$

From this, we deduce:

$$\left(\frac{s}{x}\right)^2 = \frac{K_D}{K_s} x^{n-2} \qquad [5.12]$$

Knowing the values of s and x, and thus the ratio s/x, we seek to obtain a straight line by plotting the function $\left(\frac{s}{x}\right)^2$ as a function of x^{n-2} by trying various integer values for n. When that straight line is found, we deduce n and then K_D, using relation [5.11], where we know K_s, s and x, which is the coordination index.

5.4. Solution with multiple electrolytes at equilibrium with pure solid phases

We shall now examine the equilibrium of a solution containing multiple electrolytes with one or more pure solid phases stemming from the precipitation of some of the electrolytes present in the solution.

5.4.1. *Influence of a salt with non-common ions on the solubility of a salt*

The salt under study is now added to another salt which shares no ions with the salt we are studying. The concentrations of the anion and the cation in our salt are linked to its solubility *s*, expressed in moles per liter, by the expressions:

$$\left[A^{z_+ +}\right] = \nu_+ s \qquad [5.13a]$$

and

$$\left[B^{z_- -}\right] = \nu_- s \qquad [5.13b]$$

By feeding back into expression [5.3], we obtain:

$$s = \frac{C^{onst}}{\gamma_\pm^{(\nu_+ + \nu_-)}} \qquad [5.14a]$$

where

$$C^{onst} = \left(\frac{K_s^\infty}{\nu_+^{\nu_+} \nu_-^{\nu_-}}\right)^{1/(\nu_+ + \nu_+)} \qquad [5.14b]$$

Thus, the solubility depends on the mean activity coefficient of the ions. As that activity coefficient decreases with greater ionic strength – at least for low ionic strengths, as Debye and Hückel's law demonstrates, for example – we can see that any introduction of a foreign salt with ions not shared by the salt under study, which increases the ionic strength, consequently increases the solubility *s* of our original salt. This is known as *the salt's effect on the solubility*.

By measuring the solubility *s* at different ionic strengths and extrapolating the curve $s(I)$ at an ionic strength of zero, we obtain the value of the constant C^{onst} in relation [5.14b], because then the mean activity coefficient tends toward 1, and hence, by extrapolation to an ionic strength of 0, we deduce the value of the solubility product K_s^∞. Then, on the basis of the points of the curve, we obtain the mean activity coefficient $\gamma_\pm$ at any given value of the ionic strength.

NOTE.– Looking at Debye and Hückel's model, it is advantageous to carry out the extrapolation in a coordinates system:

$$\ln s\ ,\ I^{\frac{(\nu_+ + \nu_-)}{2}}$$

to obtain a straight line at low ionic strength, which makes the extrapolation easier.

Conversely, in the conditions of application of the Debye–Hückel limit law, we can find the values K_s^∞ and s if we know the solubility s_e in pure water. Indeed, for our salt in pure water, we can write, on the basis of relation [5.14a]:

$$\ln s_e = \ln C^{onst} - \frac{(\nu_+ + \nu_-)}{2} \ln \gamma_\pm \qquad [5.15]$$

Thus, if we let I_e denote the ionic strength in pure water:

$$\ln s_e = \ln C^{onst} + \frac{(\nu_+ + \nu_-)}{2} B z_+ z_- \sqrt{I_e} \qquad [5.16]$$

Similarly, for a mixture of our salt and other electrolytes in a solution whose ionic strength is I, we have:

$$\ln s = \ln C^{onst} + \frac{(\nu_+ + \nu_-)}{2} B z_+ z_- \sqrt{I} \qquad [5.17]$$

By comparing expressions [5.16] and [5.17], we find:

$$\ln \frac{s}{s_e} = \frac{(\nu_+ + \nu_-)}{2} B z_+ z_- \left(\sqrt{I} - \sqrt{I_e}\right) \qquad [5.18]$$

In a medium of low concentration where the Debye–Hückel limit law applies fully, this relation gives us the variation in the solubility of a salt as a function of its solubility in pure water and the ionic strength.

5.4.2. *Influence of a salt with a common ion on the solubility of a salt*

When, to a saturated solution of a given salt in water, we add another salt which contains an ion $A^{z_+ +}$ that is also found in the initial salt, $A_{\nu_+}B_{\nu_-}$, we superimpose two effects:

– the effect of salt as described in the previous section 5.4.1, which leads to an increase in the solubility of our salt $A_{\nu_+}B_{\nu_-}$;

– a decrease in concentration of the ion $B^{z_- -}$ not common to the two salts, because of the existence of the solubility product constant, and therefore a decrease in the solubility of the salt $A_{\nu_+}B_{\nu_-}$.

In general, the second effect is greatly predominant.

5.4.3. *Crystallization phase diagram for a mixture of two salts in solution*

The mixture of two salts in an aqueous solution is a system containing three independent components, whose representation in a phase diagram entails the use of the triangular representation. However, in view of the special role played by water, this triangular representation actually uses a right-angled triangle rectangle with the solvent at the right angle, the concentration of each salt given on the Ox and Oy axes, and the Oz axis showing the temperature.

Figure 5.1 shows three isotherms in the crystallization diagram for the mixture water-KCl-NaCl. This diagram is used to understand the treatment of sylvinite, which is a mineral composed of the mixture of the two salts represented by the point M.

The curve $B_{100}C_{100}$ represents the line of crystallization showing sodium chloride at the temperature of 100°C, and the curve $A_{100}C_{100}$ that of crystallization of potassium chloride at the same temperature. Those curves obey relation [5.11], and we can very clearly see the effect of the common ion on the solubility of one salt or the other. The curves $B_{50}C_{50}$, $B_{10}C_{10}$ and $A_{50}C_{50}$, $A_{10}C_{10}$ have the same meanings at the temperatures of 50 and 10°C.

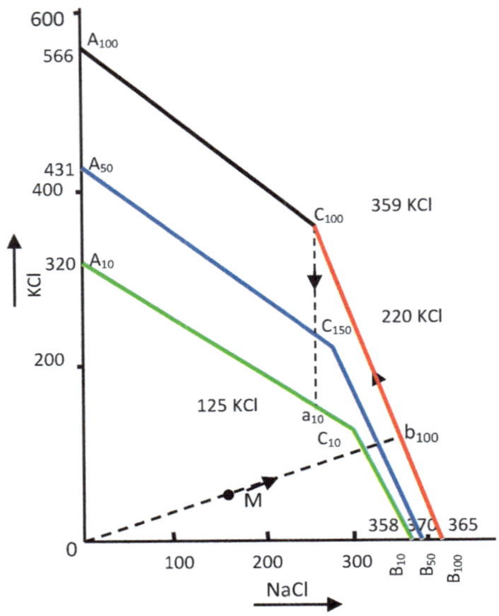

Figure 5.1. *Isotherms of crystallization of the KCl – NaCl mixture in an aqueous solution and treatment of sylvinite. For a color version of the figure, see www.iste.co.uk/soustelle/equilibria.zip*

The treatment of sylvinite consists firstly of heating the solution to 100°C; the representative point on the diagram follows the line OM to the point b_{100}. From that point on, sodium chloride begins to be deposited, and the representative point follows the curve $b_{100}C_{100}$. Just before reaching the point C_{100}, we filter out the solid sodium chloride and cool the liquid mixture to 10°C, so the representative point drops vertically along $C_{100}a_{10}$. As we are in the domain of precipitation of potassium chloride, it does indeed precipitate. Once we reach a_{10}, we filter the solution, add water and ore and start the operation again.

5.4.4. *Formation of double salts or chemical combinations in the solid state*

Evidently, like liquid–solid phase diagrams, ternary phase diagrams showing the aqueous solution and solid salts may include definite compounds such as double salts. This results in the appearance on the diagram of a curve representing the precipitation of that new phase.

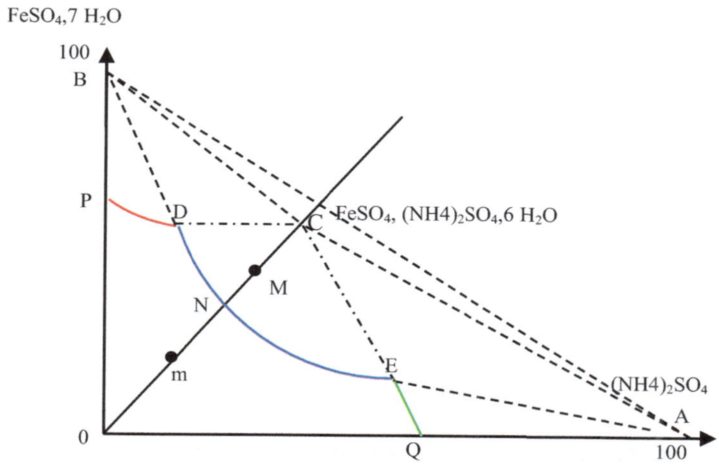

Figure 5.2. *Crystallization diagram of the mixture of iron sulfate and ammonium sulfate with the formation of Mohr's salt. For a color version of the figure, see www.iste.co.uk/soustelle/equilibria.zip*

Figure 5.2 shows the example of the diagram for the system of water, ferrous sulfate heptahydrate ($FeSO_4,7H_2O$)–ammonium sulfate. These last two salts may give rise to a definite (a mixed salt or double salt) known as Mohr's salt, which has the formula $FeSO_4,(NH_4)_2SO_4,6H_2O$.

In Figure 5.2, the precipitation of that mixed salt is shown by the apparition of the branch DE of the saturation of Mohr's salt.

In certain cases, unlike the one discussed above, the extension of the line Om by evaporation does not end in the domain of precipitation CDE of the double salt. Such is the case, represented by Figure 5.3, of the magnesium hexahydrate chloride ($MgCl_2,6H_2O$)–potassium chloride system. These two salts give rise to a double salt: carnallite, whose formula is $MgCl_2,KCl,6H_2O$. Figure 5.3 shows that a solution represented by the point m ultimately evaporates at point m', intersecting the curve EQ of precipitation of potassium chloride at point N. Thus, carnallite has incongruent dissolution, whereas Mohr's salt, which we saw above, had congruent dissolution. To obtain carnallite by evaporation, we would have had to start with a solution richer in magnesium sulfate, whose representative point would be situated, for instance, on the line OK. That line would then intersect the branch DE of carnallite precipitation at point N'.

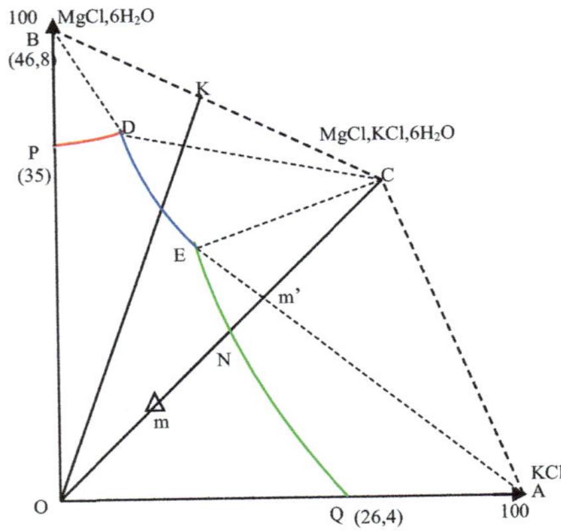

Figure 5.3. *Diagram of crystallization of carnallite. For a color version of the figure, see www.iste.co.uk/soustelle/equilibria.zip*

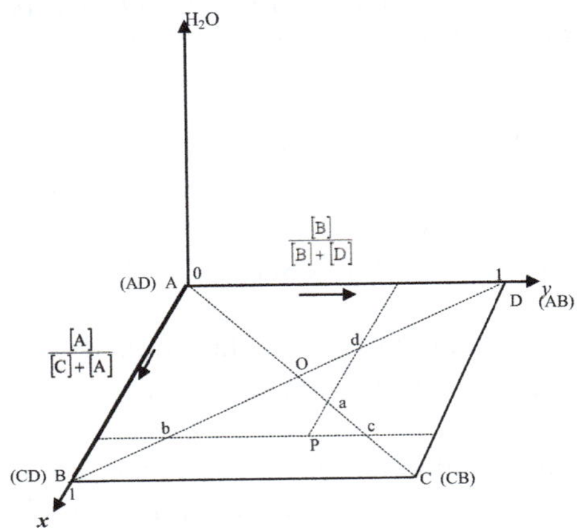

Figure 5.4. *Square representation of reciprocal quaternary diagrams*

5.4.5. *Reciprocal quaternary systems – square diagrams*

We now consider a system formed of water and four salts containing a common ion, two by two: AD, AB, CB and CD, between which the following reaction takes place in solution:

$$AD + CB = CD + AB \qquad [5R.3]$$

This system, known as a *reciprocal quaternary system*, contains four independent components (five components – the water and the four salts, with a defined relation between them). We can represent any given composition of that system by a point placed in a Cartesian system of axes, with the abscissa showing the cationic fraction $\frac{[C]}{[C]+[A]}$ and the ordinate showing the anionic fraction $\frac{[B]}{[B]+[D]}$, and on the z axis, we write the mass fraction of water – i.e. the mass of water corresponding to a given mass of mixture. Thus, we obtain a square diagram of crystallization (Figure 5.4), wherein each vertex of the square represents a pure salt: AD, CD, CB and AB. We can easily show that any point within that square – say, the point P – corresponds to a mixture which can be made up using only three salts:

– the salts AD, CD and CB. For each, we take a proportional quantity in respect to the lengths of the segments aC, ac and aA;

– the salts CD, CB and AB. For each, we take a proportional quantity in respect to the lengths of the segments dD, bd and bB.

To create the composition represented by the center O of the square denoting the mixture, equal quantities of two salts will suffice: either AD and CB or AB and CD.

To illustrate this type of diagram, we have chosen to use the system composed of water, sodium chloride (NaCl), ammonium chloride (NH_4Cl), sodium hydrogen carbonate ($NaHCO_3$) and ammonium hydrogen carbonate (NH_4HCO_5). This system is the basis from which we obtain solid sodium hydrogen carbonate by the Solvay process, with the four salts being linked by the reaction:

$$NH_4HCO_3 + NaCl = NaHCO_3 + NH_4Cl \qquad [5R.4]$$

Figure 5.5 shows the square diagram of that system at the temperature of 15°C.

When, because of the evaporation of the water, a solid salt appears, the system contains two phases (liquid and solid), and its variance becomes equal to 5. If we set the temperature, the composition of the solution (point in the square) and the mass fraction of water, the system then becomes part of a crystallization surface formed of four layers, each surrounding one of the vertices of the square, with very different values. It is usual to represent these surfaces by their level lines, projected onto the square. Each line shows the corresponding mass of water per mole of mixture (see Figure 5.5).

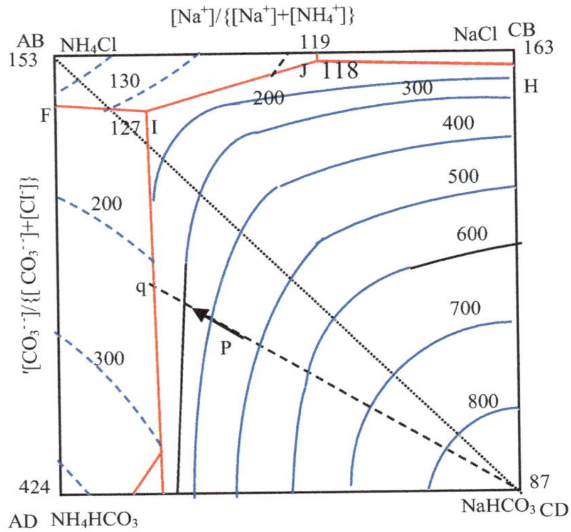

Figure 5.5. *Square diagram of the system NaCl, NH_4HCO_3, H_2O. For a color version of the figure, see www.iste.co.uk/soustelle/equilibria.zip*

The crystallization layers intersect following lines along which we see the crystallization of a mixture of two salts, and there are two triple points, projected at points I and J on the square, where three salts crystallize simultaneously. The position of those points depends solely on the temperature.

If, at 15°C, we evaporate a solution represented by the point P in Figure 5.5, in the domain of sodium hydrogen carbonate, the ratio between the quantities of the chlorine ion and the ammonium ion is not altered by the deposition of sodium hydrogen carbonate. The point in the square diagram representing the liquid's composition moves away from the vertex corresponding to the salt that has been deposited, along the line (CD)P. When the composition reaches q, ammonium hydrogen carbonate begins to be deposited, and the point representing the solution's composition moves along the line qJ, with a mixture of the two hydrogen carbonates being deposited. To obtain sodium hydrogen carbonate, therefore, we need to stop just before point q.

5.5. Electrolytic aqueous solution and solid solution

It may happen, in certain cases, that the two solids in the liquid–solid binary system are miscible with one another, producing mixed crystals which form a solid solution.

5.5.1. *Thermodynamic equilibrium between a liquid ionic solution and a solid solution*

If the two solids are miscible in all proportions, we say that we have a simple isomorphism.

The variance of the equilibrium between the liquid and solid solutions is 2; keeping the temperature constant, we need only fix the composition of one solution, and that of the other solution is fixed too as a result. In other words, at equilibrium, there is a relation which exists between the compositions of the two solutions.

Figure 5.6(a) shows the example of the crystallization diagram of the system $CuSO_4$, 5 H_2O-$MnSO_4$, 5 H_2O above 10°C. Given what we have just seen, there is only one crystallization curve. At the composition of the aqueous phase represented by point K, for instance, at equilibrium, there is a corresponding composition of the solid phase, given by point N, which is on the straight line AB. Typically, the addition of one of the two salts to the mixture enriches the aqueous solution more quickly than the solid solution, which explains why the configuration lines, such as KN, which are almost parallel to the axes when near to those axes, quickly change direction in the middle region.

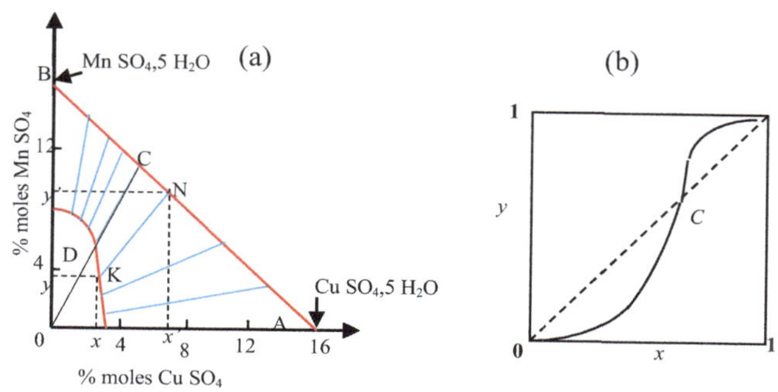

Figure 5.6. *a) Phase diagram for a liquid–solid solution with simple isomorphism and b) concentration/concentration curve. For a color version of the figure, see www.iste.co.uk/soustelle/equilibria.zip*

To establish the relation between the compositions of the two solutions, we must first note that in each of them, because of the law of electrical neutrality, the quantity of sulfate is equal to the sum of the quantities of copper and manganese:

$$n_{Cu} + n_{Mn} = n_{SO_4} \qquad [5.19]$$

The composition of the liquid solution is measured by the cationic fraction of copper:

$$x^{(\mathrm{L})} = \frac{n_{Cu}}{n_{Cu} + nMn} \qquad [5.20]$$

The activity coefficients of the copper and manganese ions are, respectively, $\gamma_{\mathrm{Cu}}^{(\mathrm{L})}$ and $\gamma_{\mathrm{Mn}}^{(\mathrm{L})}$.

The composition of the solid solution is measured by the molar fraction $x^{(\mathrm{S})}$ of copper sulfate pentahydrate, with activity coefficients $\gamma_{\mathrm{Cu}}^{(\mathrm{S})}$ and $\gamma_{\mathrm{Mn}}^{(\mathrm{S})}$.

The equalities of the chemical potentials of copper and manganese in the two solutions give us:

$$\frac{x^{(\mathrm{L})}\gamma_{\mathrm{Cu}}^{(\mathrm{L})}}{x^{(\mathrm{S})}\gamma_{\mathrm{Cu}}^{(\mathrm{S})}} = K_{\mathrm{Cu}} \qquad [5.21a]$$

and

$$\frac{\left(1 - x^{(\mathrm{L})}\right)\gamma_{\mathrm{Mn}}^{(\mathrm{L})}}{\left(1 - x^{(\mathrm{S})}\right)\gamma_{\mathrm{Mn}}^{(\mathrm{S})}} = K_{\mathrm{Mn}} \qquad [5.21b]$$

The ratio between these two equations gives us a sharing coefficient:

$$\frac{x^{(\mathrm{L})}(1 - x^{(\mathrm{S})})\gamma_{\mathrm{Cu}}^{(\mathrm{L})}\gamma_{\mathrm{Mn}}^{(\mathrm{S})}}{x^{(\mathrm{S})}(1 - x^{(\mathrm{L})})\gamma_{\mathrm{Mn}}^{(\mathrm{L})}\gamma_{\mathrm{Cu}}^{(\mathrm{S})}} = \frac{K_{\mathrm{Cu}}}{K_{\mathrm{Mn}}} = K_{part} \qquad [5.22]$$

Figure 5.6(b) shows the curve of variation of the composition of the solid phase as a function of the composition of the liquid phase. This curve intersects the bisector at point C. It is easy to show that fractional crystallization ultimately yields a pure salt (that which is richest initially), and the mixture of the two salts corresponding to point C. Indeed, at that point, the solid which crystallizes has the same composition as the liquid solution, which prevents further enrichment.

If the two salts are not miscible in all proportions, we obtain two solid solutions, respectively rich in one and the other of the components. We then say that we have isodimorphism. Figure 5.7(a) shows the example of such a situation for the system $CuSO_4$, $5H_2O$-$MnSO_4$, $7H_2O$ at a temperature less than 10°C.

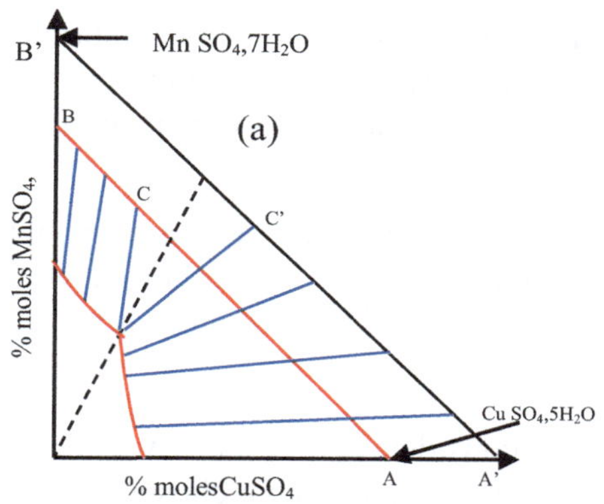

Figure 5.7. *a) Liquid-solid solution phase diagram with isodimorphism and b) concentration/concentration curve. For a color version of the figure, see www.iste.co.uk/soustelle/equilibria.zip*

The crystallization diagram then exhibits two branches depending on which of the two solid solutions is precipitated. The point C is on the line AB, which connects manganese sulfate heptahydrate to a hypothetical copper sulfate heptahydrate. Meanwhile, C' is on the line A'B', joining copper sulfate pentahydrate to a hypothetical manganese sulfate pentahydrate.

The composition curve shown in Figure 5.7(b) includes two branches – OD and O'D' – which a gap between them which corresponds to the domain CC' in the phase diagram 5.7(a).

5.5.2. *Solubility product of a solid solution*

Consider a solid solution $X_xY_{1-x}A$, whose two pure poles have the formula AX and BY. That solid solution dissolved in water to give cations X^{z+} and Y^{z+} and anions A^{z-}. To simplify the writing, we choose $z = 1$, allowing us to reason freely. At the temperatures at which dissolution in water takes place, there is no diffusion of the ions in the solid state, so no exchange between the ions X and Y within the solid. This means that the thermodynamic equilibrium of the solution as expressed by the equality of the chemical potentials can never be reached in the solid. The state of equilibrium between the liquid and solid solution, therefore, can be dealt with in two different ways:

– the first consists of writing that only the surface of the solid solution in contact with the liquid solution is at equilibrium with it, but then that equilibrium must involve the properties of the surface, whose contribution to the Helmholtz energy must not be overlooked;

– the second solution was developed by Thorstenson and Plummer, who introduced the notion of stoichiometric saturation. It is this technique which led to the concept of the *solubility product of a solid solution*, which we have chosen to look at.

In order to develop the model, we consider two salts XA and YA, composed of one anion A^- and cations X^+ and Y^+. We have chosen 1:1 salts to simplify the expressions, but it is perfectly easy to switch to other types. Thus, we have a solid which, in solution, contains those two salts with the molar fractions $x^{(S)}$ for the salt XA and $1-x^{(S)}$ for the salt YA. Note that $x^{(S)}$ therefore represents the cationic fraction of X and $1- x^{(S)}$ the cationic fraction of Y.

This salt is placed in water, in which it gradually dissolves, yielding a dilute solution where the molar fraction of the salt is X, and that of the water is $1-X$. In the course of this dissolution, the composition of the solid is not altered; the mixed salt is said to show congruent dissolution.

Its dissolution in water can be represented by the following balance equation:

$$<(X+Y)A> = (X^+) + (Y^+) + (A^-) \qquad [5R.5]$$

In modeling the constant composition of the solid, we considered that the cations formed a coherent set (X+Y), whose activity coefficient is written as $\gamma^{(S)}_{(X+Y)}$.

Table 5.1 gives the molar fractions and activity coefficients of the various species.

NOTE.– The term ½ of the molar fraction of A in the solid stems from the fact that we have chosen two salts of the same type – here 1:1 – for our example.

We shall first base our discussion on the hypothesis of a perfect solution with activity coefficients of 1. We shall then look at the case of non-perfect solutions, and see the hypotheses that need to be made in such a case.

	x_A	x_X	x_Y	γ_A	γ_X	γ_Y
Liquid	$X/2$	$x^{(L)}X/2$	$(1-x^{(L)})X/2$	$\gamma_A^{(L)}$	$\gamma_X^{(L)}$	$\gamma_Y^{(L)}$
Solid	1/2	$x^{(S)}/2$	$(1-x^{(S)})/2$	$\gamma_A^{(S)}$	$\gamma_X^{(S)}$	$\gamma_Y^{(S)}$

Table 5.1. *Notations used in the solubilization of a solid solution in water*

At saturation, i.e. with the molar fractions shown in Table 5.1, the quotient of reaction [5R.11] is written:

$$Q_{sat} = \frac{\left[x^{(L)}X + \left(1-x^{(L)}\right)X\right]X}{x^{(S)} + \left(1-x^{(S)}\right)} \qquad [5.23]$$

This expression gives us an infinite number of states, but it is important that it be valid at the limits – i.e. for the pure compound XA, which dissolves as follows:

$$\langle XA \rangle = (X) + (A) \qquad [5R.6]$$

Even if the molar fraction is 1, we have its solubility product such that:

$$K_{XA} = \frac{x^{(L)}X^2}{x^{(S)}} \qquad [5.24a]$$

Similarly, for the pure compound YA:

$$K_{\mathrm{YA}} = \frac{\left(1 - x^{(\mathrm{L})}\right) X^2}{1 - x^{(\mathrm{S})}} \qquad [5.24b]$$

By feeding equations [5.24a] and [5.24b] back into relation [5.15], we find a particular value K_{sat} of the quotient Q_{sat}, given by:

$$Q_{sat}^{(equ)} = K_{sat} = K_{\mathrm{XA}} x^{(\mathrm{S})} + K_{\mathrm{YA}} \left(1 - x^{(\mathrm{S})}\right) \qquad [5.25]$$

The constant K_{sat} plays the role of a total solubility product of the solid solution, which appears as the sum of the partial solubility products of each of the poles of the solid solution, weighted by the molar fraction of the corresponding pole in the solid.

Note the similarity between expression [5.25] and the relation giving the total vapor pressure above a perfect liquid solution of two components as a function of the saturating vapor pressures of those pure components; the total solubility product K_{sat} plays the role of the total pressure, whilst the two solubility coefficients K_{XA} and K_{YA} of the poles play the roles of the partial pressures.

We shall now look at the calculations in the case of non-perfect solid solutions and liquids. The calculation method is the same as above, except that we now take account of the activity coefficients. The solubility relation [5.23] is replaced by the expression:

$$Q_{sat} = \frac{\left[x^{(\mathrm{L})} \gamma_{\mathrm{X}}^{(\mathrm{L})} + \left(1 - x^{(\mathrm{L})}\right) \gamma_{\mathrm{Y}}^{(L)}\right] \gamma_{\mathrm{A}}^{(\mathrm{L})} X^2}{\gamma_{(\mathrm{X+Y})}^{(\mathrm{S})} \left[x^{(\mathrm{S})} + \left(1 - x^{(\mathrm{S})}\right)\right]} \qquad [5.26]$$

The solubility products of the two pure poles are, respectively:

$$K_{\mathrm{XA}} = \frac{x^{(\mathrm{L})} X^2 \gamma_{\mathrm{A}}^{(\mathrm{L})} \gamma_{\mathrm{X}}^{(\mathrm{L})}}{x^{(\mathrm{S})} \gamma_{\mathrm{X}}^{(\mathrm{S})}} \qquad [5.27]$$

and

$$K_{\mathrm{YA}} = \frac{\left(1 - x^{(\mathrm{L})}\right) X^2 \gamma_{\mathrm{A}}^{(\mathrm{L})} \gamma_{\mathrm{Y}}^{(\mathrm{L})}}{\left(1 - x^{(\mathrm{S})}\right) \gamma_{\mathrm{Y}}^{(\mathrm{S})}} \qquad [5.28]$$

These expressions replace formulae [5.24a] and [5.24b]. Expression [5.25] then becomes:

$$Q_{sat}^{(equ)} = K_{sat} = \frac{K_{\mathrm{XA}} x^{(\mathrm{S})} \gamma_{\mathrm{X}}^{(\mathrm{S})}}{\gamma_{(\mathrm{X+Y})}^{(\mathrm{S})}} + \frac{K_{\mathrm{YA}} \left(1 - x^{(\mathrm{S})}\right) \gamma_{\mathrm{Y}}^{(\mathrm{S})}}{\gamma_{(\mathrm{X+Y})}^{(\mathrm{S})}} \quad [5.29]$$

If we want to preserve the similarity with the expression of the total vapor pressure above a solution, which is written:

$$Q_{sat}^{(equ)} = K_{sat} = K_{\mathrm{XA}} a_{\mathrm{X}}^{(\mathrm{S})} + K_{\mathrm{YA}} a_{\mathrm{Y}}^{(\mathrm{S})} \quad [5.30]$$

we are forced to set:

$$\gamma_{(\mathrm{X+Y})}^{(\mathrm{S})} = 1 \quad [5.31]$$

This mirrors the hypothesis adopted in the model, which consisted of considering the solid solution, with invariable composition, as a pure solid.

This relation was first established by Lippmann.

Thus, we simultaneously employ Thorstenson and Plummer's hypothesis – describing the behavior of the solid solution as being similar to that of a pure compound of the same composition, and where that total activity was equal to 1 – and Gresens' hypothesis, because in the denominator in relation [5.26], the total activity of the solid phase is defined as the sum:

$$a_{\mathrm{XYA}}^{(\mathrm{S})} = a_{\mathrm{XA}}^{(\mathrm{S})} + a_{\mathrm{YA}}^{(\mathrm{S})} \quad [5.32]$$

NOTE.– Owing to the need to respect electrical neutrality and preserve the solid's composition, we obviously have congruence on dissolution, which is expressed by the equality:

$$x^{(\mathrm{L})} = x^{(\mathrm{S})} \quad [5.33]$$

From the similarity with the vapor pressures of the solutions, mentioned above, we deduce a graphical representation which shows an equilibrium between a liquid solution and a solid solution, in the same way as isothermal curves do for liquid–vapor equilibria.

Figure 5.8 shows the plot in the case of total miscibility between the two solids.

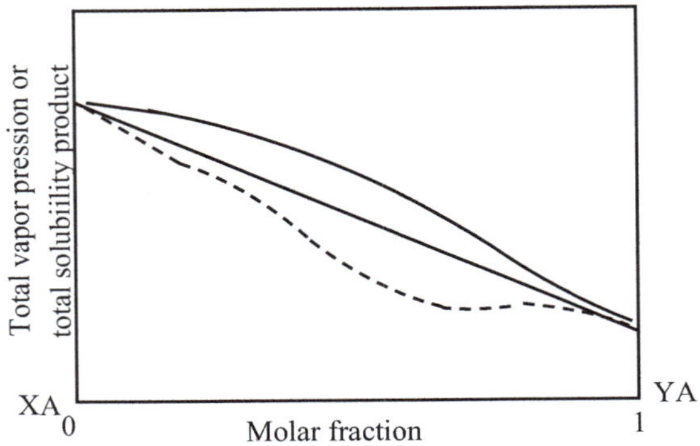

Figure 5.8. *Analogy between the pressure/composition diagram for a liquid–vapor equilibrium and the total solubility product of a system comprising two solutions: liquid and solid*

We may find alyotropic compositions, which are compositions equivalent to the azeotropic compositions – i.e. where the two curves of the solidus (composition of the solid solution) and the liquidus (corresponding composition of the liquid solution) intersect, and present a common extremum. Figure 5.9 illustrates such a case for the system $FeCO_3$-$MnCO_3$–H_2O, according to Lippmann.

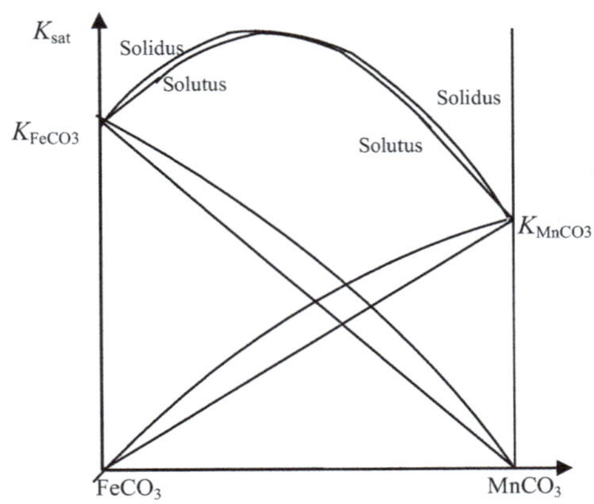

Figure 5.9. *Diagram with alyotropic maximum of the system $FeCO_3$-$MnCO_3$-H_2O (according to Lippmann)*

NOTE.– In relation [5.15], the constants K_{XA} and K_{YA} are the solubility products of the pure salts XA and YA. This is accurate only in the case of total miscibility of the two salts in the solid phase, because then they have the same crystallographic structure. In the case of partial miscibility, we need to choose which of the two salts imposes its crystallographic structure on the whole solid solution – generally, it will be that which is more abundant. Thus, the solubility product of that salt to be fed into relation [5.30] will indeed be its true solubility product, but the constant to be entered for the salt in to minority will be a pseudo-solubility product – i.e. the solubility product that the second solid would have if it were to crystallize in the pure state in the crystalline system of the majority salt.

Obviously, the calculations presented above need to be repeated in the case of electrolytes which do not have a 1:1 structure.

5.6. Solubility and pH

The pH or pX of a solution may influence the solubility of certain species.

5.6.1. *Solubility and pH*

In a certain number of cases, the solubility of a cation in a protic solvent may be dependent upon the pH. This phenomenon is observed, for example, when a hydroxide is insoluble and amphoteric, meaning that it behaves like an acid when interacting with a stronger base, and like a base when interacting with a stronger acid.

Such is the case, for instance, with zinc hydroxide and aluminum hydroxide.

Let us take the example of an aqueous solution of aluminum chloride in an acidic medium. If we increase the pH by adding a base – caustic soda, for instance – then the aluminum hydroxide precipitates. If we continue adding caustic soda, the aluminum re-dissolves, giving aluminate ions AlO_2^-.

In fact, the aluminum hydroxide, which is amphoteric, participates in two acid–base couples:

– the aluminum ion/hydroxide couple, which works as follows:

$$Al^{3+} + 6H_2O = Al(OH)_3 + 3H_3O^+ \qquad [5R.7]$$

– the hydroxide/aluminate ion couple, as follows:

$$Al(OH)_3 + H_2O = Al(OH)_4^- + H_3O^+ \qquad [5R.8]$$

The pK of the acids are respectively 4.3 and 10.5.

The diagram showing the domains of predominance (Figure 5.10) is easy to plot.

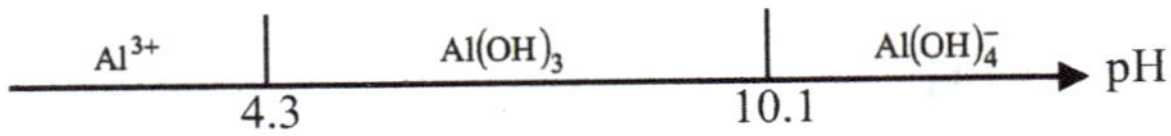

Figure 5.10. *Domains of predominance of the species deriving from the aluminum ion as a function of the pH*

Thus, aluminum is soluble at pH levels lower than 4.3 or higher than 10.1, but is insoluble between those two values.

5.6.2. *Solubility of oxides in molten alkali hydroxides*

The solubility of metal oxides is generally very low in molten alkali hydroxides. A metal oxide, therefore, may precipitate when an acidic solution – in our case, a hydrated solution – of the corresponding hydroxide is neutralized. That neutralization may be achieved either by the addition of alkali oxides or by the elimination of water vapor. We have the reaction:

$$M^{+n} + nOH^- = MO_{n/2}(sol) + \frac{n}{2}H_2O \qquad [5R.9]$$

Certain oxides are amphoteric, and may be re-dissolved by an increase in the pH_2O.

Here, we are seeing properties that absolutely mirror those of hydroxides in aqueous media, as described in section 5.6.1. Water, with molten salts, plays the same role as the proton plays with dissolved salts.

5.6.3. *Solubility in oxo-acids and oxo-bases (see section 3.12.2)*

Consider the equilibrium of dissolution of a solid in a solvent such as the molten KCl-LiCl mixture. It can be written in the form:

$$M_aO_b = aM^{n+} + bO^{2-} \qquad [5R.10]$$

That solubility can be characterized by the solubility product:

$$K_S = \left[M^{n+}\right]^a \left[O^{2-}\right]^b \qquad [5.34]$$

We can define the solubility by the value of the pO^{2-} corresponding to the saturation for a concentration of ions M^{n+} – say, 1mol/l. We can then write:

$$pO^{2-} = \frac{pK_S}{b} \qquad [5.35]$$

The dissolution of an oxide can be represented as an oxo-acid–base reaction such as:

$$< ZnO > + Ni^{2+} = < NiO > + Zn^{2+} \qquad [5R.11]$$

Figure 5.11, on a pO^{2-} scale, shows firstly the portion of pO^{2-} of precipitation of a certain number of oxides (top scale) for a cationic concentration of 1mol/l, and secondly the pK_a of acid–base couples (bottom scale).

On this graph, we can see, for example, that if we gradually introduce carbon dioxide into the solution, it is only possible to dissolve zinc oxide and cobalt oxide. On the other hand, using the acid PO_3^-, we can dissolve all the oxides mentioned, with the exception of silica.

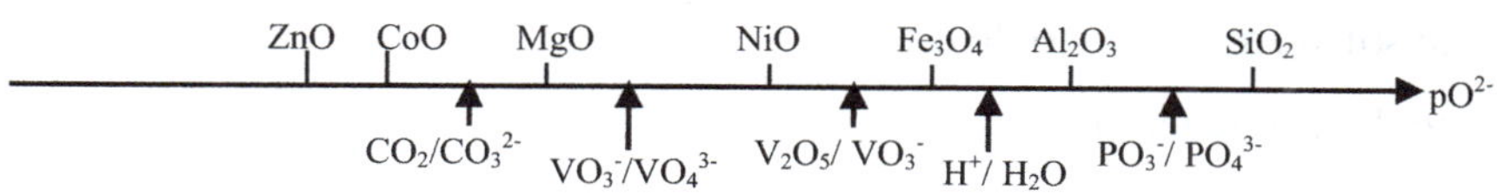

Figure 5.11. *Dissolution of oxides by oxo-acid–base reactions (from [GUE 09])*

5.7. Calculation of equilibria in ionic solutions

Having reviewed the different ionic exchange reactions in solution, we now need to look at methods for calculating a system's state of equilibrium on the basis of the amounts of materials constituting it and the characteristics of the various equilibria which are likely to occur.

There are two main categories of methods to solve ionic systems at equilibrium. These systems involve the superposition of several equilibria.

We shall not linger over manual approximation methods which can be used to solve simple systems containing one or two equilibria; readers will undoubtedly already be familiar with these methods. Let us simply state that, by way of certain approximations, they generally yield easy-to-use analytical expressions. The most practical and systematic method is the predominant reaction method, looking at the disjointed domains of predominance of each of the species. This method is used only in very dilute solutions, so as to be able to treat the concentrations and activities as being the same thing. Readers wanting further details or a re-examination of this method can refer to the excellent book by Bernache-Assolant and Cournil [BER 97].

We shall look in greater depth at automated calculation methods, which usually rely on a database giving the equilibrium constants. Remember that usually, it is a question of establishing the necessary equilibria to find the concentrations of the different species of a system placed in the specified conditions. This is known as the *speciation* of the species.

By way of example, we shall describe *PhreeqC*: a geochemical computation tool for aqueous solutions. That software has already yielded excellent results, and has the advantage of being accessible, freely, to download. Version 3 can be obtained from *wwwbrr.cr.usgs.gov/projects/GWC.../phreeqc/*.

The software is able to deal with:

– acid–base reactions;

– redox couples;

– complexation reactions;

– precipitation reactions;

– the different phase equilibria.

The computational principle is the same as that in most automated calculation software tools, and is represented by the workflow in Figure 5.11.

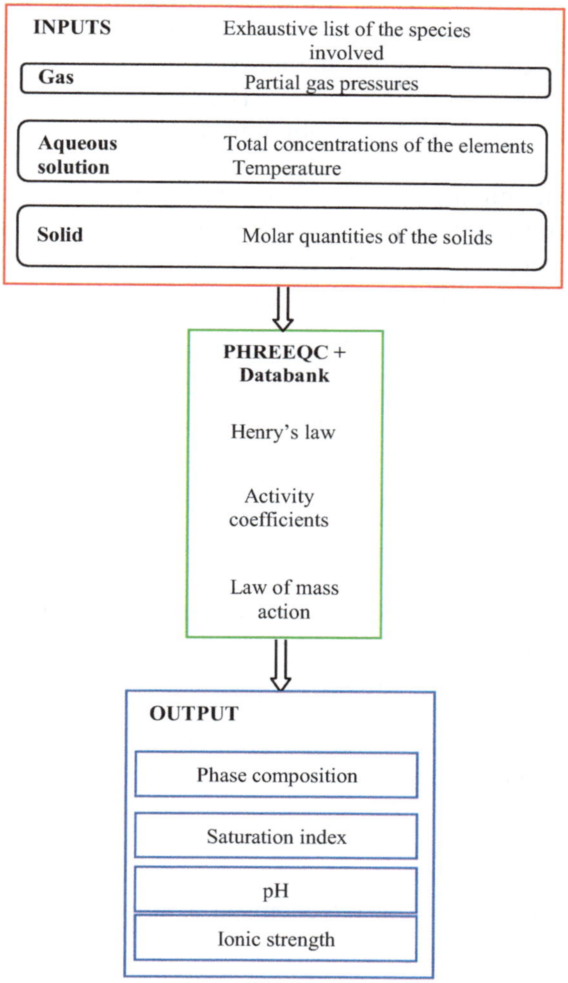

Figure 5.12. *Principle behind pHREEQc*

This figure shows the three blocks: the input block, the processing block by the program and the output block.

Obviously, it is essential to input *all* the compounds which play a part. Users must, of course, list all the initial products, but also all of the possible final products, even if some of them will not ultimately be encountered in the list of species at equilibrium. If a species mentioned is not present in the final state, this means that the species is not stable in the chosen conditions, which is completely different to the absence, from the end products, of a species which was never mentioned. The result will never include the concentration of a species which is not indicated.

For gases, the initial partial pressures are input. For solutions, we input the dissolved concentrations, whether the materials were originally in the solid or liquid state. Of course, we also need to set the temperature at which we wish to achieve equilibrium.

The system works using a database. There are many such databases available, and the user must indicate which s/he wishes to draw upon. The program is able to modify the databases and add to them with components that are absent and not filled in.

We then input a certain number of instructions, which are requested by the program and are used to simulate the reactions involved.

The program works for closed systems. It uses two families of equations: the equilibrium equations between the activities a_i and the material balance for each chemical element involved.

With regard to the equilibria, the program uses Henry's law for gas–liquid equilibria, written in the form:

$$P_i = K_{iH} a_i \qquad [5.36]$$

The law of mass action for the chemical equilibria or phase equilibria is used, written in the form:

$$K_R = a_i \prod_{j \neq i}^{N} a_j^{\nu_{j(R)}} \qquad [5.37]$$

In these expressions, the stoichiometric coefficients pertaining to relation R and to the components j: $\nu_{j(R)}$ are expressed in terms of one mole for species i in the writing of the reaction.

These equilibrium laws require us to calculate the activities of the different species. They are calculated as follows:

– for pure solid substances, the activity is equal to 1;

– for solids in a solid solution, the activities are replaced by the concentrations;

– for liquid solutions, the activities are calculated by the product of the concentration by an activity coefficient, which is determined by one of the following equations – each used in a particular range of values of the ionic strength;

- if the ionic strength I is less than 0.05 mol/l, we use the Debye–Hückel relation:

$$\ln \gamma_i = -Az_i^2 \frac{\sqrt{I}}{1+\sqrt{I}} \quad [5.38]$$

- if the ionic strength I is less than 0.5 mol/l, we use the Davies equation:

$$\ln \gamma_i = -Az_i^2 \left(\frac{\sqrt{I}}{1+\sqrt{I}} - 0{,}3I \right) \quad [5.39]$$

- if the ionic strength I is less than 1 mol/l, we use an extended Debye–Hückel relation:

$$\ln \gamma_i = -Az_i^2 \left(\frac{\sqrt{I}}{1+rB\sqrt{I}} - BI \right) \quad [5.40]$$

In these expressions, z_i is the charge of the ion i, I is the ionic strength of the medium, A and B are constants dependent only on the temperature (hence the need to state the temperature), and r is an apparent radius of the ions.

The second category of equations is formed of material balances for all the elements introduced.

If no other specification is input, the program will automatically use the equation of electrical neutrality. That equation can be replaced by inputting an imposed concentration of a species – e.g. the pH of the final solution, or the ionic strength.

As output, the program gives the speciations of each of the species, and if solids are involved, the program is able to calculate the saturation index *SI*, defined on the basis of the solubility product and the reaction quotient of the precipitation reaction, by the relation:

$$SI = \ln\left(\frac{K_S}{Q_S}\right) \qquad [5.41]$$

Thus, we see whether the solution is under-saturated ($SI < 0$), at equilibrium ($SI = 0$) or oversaturated ($SI > 0$). Equilibrium may not necessarily be achieved if it is incompatible with the input constraints (e.g. the pH).

Of course, the quality of the results obtained depends on the accuracy of the input and the quality of the data gleaned from the databases.

There are other commercially-available tools, but as before, the quality of the results depends heavily on the quality of the data.

PART 2

Electrochemical Thermodynamics

6

Thermodynamics of the Electrode

In this chapter, we examine the thermodynamics of so-called *electrochemical* reactions and equilibria which produce or consume electrons.

These reactions take place on contact with electron-conductive electrodes, and in studying them, we employ electrochemical thermodynamics.

6.1. Electrochemical systems

In keeping with J. Besson [BES 84], we shall use the term "*electrochemical system*" to denote any mono- or polyphasic system involving electrostatic forces sufficiently strong to influence the energy exchanges in that system. This happens if the electrical field is not null everywhere in the system – put differently, if that system contains active charges.

NOTE.– The reactions in ionic solutions which we looked at in the previous chapters are not electrochemical systems, because the media in which they occur do not contain any active charges, owing to the imposition of local electrical neutrality at every point in the system.

6.1.1. *The electrochemical system*

Thermodynamics of electrochemical systems will be a transposition of conventional thermodynamics[1] with the involvement of electrical energy, so the chemical values are replaced by electrochemical ones.

When the system is at equilibrium, the active charges are localized on inter-phase surfaces where the electrical field is perpendicular to the inter-phase surface.

Thus, besides the composition variables, an electrochemical system involves three couples of intensive and extensive values:

– temperature–entropy;

– pressure–volume;

– electrical field–electrical charges.

As the electrochemical system is always supposed to be at thermomechanical equilibrium, the existence of an electrical field normal to the interfaces causes what is known as electrostatic pressure in each phase. These pressures balance one another out, and therefore the pressure involved in thermodynamic expressions is always the pressure outside of the system.

An electrochemical reaction is a heterogeneous reaction which takes place at the interface between two media, which amounts to a modification of the chemical- and electrical energies of those two media. Very often, the electrochemical reaction involves the changing of the mode of electricity conduction from one medium to another. For instance, at the interface between a metal and an aqueous solution, we shift from electronic conduction in the metal to ionic conduction in the solution. At the interface between a semi-conductor n and a semi-conductor p, the system transitions from conduction by electrons in the semi-conductor n to conduction by electron holes in the semi-conductor p.

The electrode constitutes an important specific case of an electrochemical system.

---

1 In reading this section, readers should refer to Volume 1 of this collection: *Phase Modeling Tools* [SOU 15a].

6.1.2. *Electrochemical functions of state*

We shall examine the thermodynamic change caused by the intervention of electrical work.

By including, in the work W, the work of the electrical forces as well as that of the forces of pressure and heat energy, we define the following differential of the *internal electrochemical energy*:

$$\mathrm{d}\tilde{U} = T\,\mathrm{d}\tilde{S} - P\,\mathrm{d}V + \Phi\,\mathrm{d}q \qquad [6.1]$$

We then define the various characteristic functions, such as the *electrochemical enthalpy*:

$$\tilde{H} = U + PV + \Phi q = \tilde{U} + PV = H + \Phi q \qquad [6.2]$$

the *electrochemical Helmholtz energy*:

$$\tilde{F} = U - T\tilde{S} + \Phi q = \tilde{U} - T\tilde{S} \qquad [6.3]$$

and the *electrochemical Gibbs energy*:

$$\tilde{G} = U + PV - T\tilde{S} + \Phi q = \tilde{U} + PV - T\tilde{S} \qquad [6.4]$$

All of the above are functions of state.

NOTE.– In the last two expressions, we included only an electrochemical entropy that is different to the entropy in the absence of a field. Indeed, whilst we can agree that the volume of a phase is not influenced by electrical field – particularly when we are dealing with condensed phases – the entropy in the absence of a field may not necessarily be identical to that in the presence of electrical forces, because the different arrangements of the species may be influenced by those electrical forces.

6.1.3. *Electrochemical potential*

The electrochemical potential is a generalized chemical potential introduced previously (see Volume 1 of this set [SOU 15a]): that with which, at the energies involved in chemical systems, we adjoin the electrical energy.

6.1.3.1. *Definition*

The *electrochemical potential* is the partial molar electrochemical Gibbs energy defined by the relation:

$$\tilde{\mu}_i = \left(\frac{\partial \tilde{G}}{\partial n_i}\right)_{P,T,n_j} = \bar{\tilde{G}}_i \quad [6.5]$$

The term $\bar{\tilde{G}}_i$ is the *partial molar electrochemical Gibbs energy* of the component i.

6.1.3.2. *Variations of the electrochemical potential with the intensive variables*

The different variables envisaged are, firstly, the intensive variables – the pressure, temperature and electrical potential – and secondly, the quantities of matter in the phase in question.

To find the variations of the electrochemical potential with the intensive variables, we use the general relation applied to electrochemical systems, so:

$$\frac{\partial \tilde{\mu}_i}{\partial Y} = -\bar{\tilde{X}}_i \quad [6.6]$$

In this expression, Y is an intensive value, meaning that in the case of interest to us here, the temperature, the opposite of the pressure or the electrical field and $\bar{\tilde{X}}_{k_i}$ is the electrochemical partial molar value of the component i of the variable X_k, which is an extensive value, the conjugate of Y_k (here the entropy, volume or electrical charge).

6.1.3.3. *Helmholtz's relations for electrochemical systems*

Let us calculate the expression $\dfrac{\partial\left(\tilde{G}/T\right)}{\partial T}$.

$$\frac{\partial\left(\tilde{G}/T\right)}{\partial T} = \frac{1}{T}\frac{\partial \tilde{G}}{\partial T} - \frac{\tilde{G}}{T^2} = \frac{\tilde{S}}{T} - \frac{\tilde{G}}{T^2} \quad [6.7]$$

Given relations [6.2] and [6.4], we have:

$$\tilde{H} = \tilde{G} + T\tilde{S} \qquad [6.8]$$

By comparison, we deduce Helmholtz's first equation:

$$\frac{\partial\left(\tilde{G}/T\right)}{\partial T} = -\frac{\tilde{H}}{T^2} \qquad [6.9]$$

By differentiating both sides of expression [6.9] in relation to the quantity n_i of component i, we establish Helmholtz's second relation:

$$\frac{\partial\left(\tilde{\mu}_i/T\right)}{\partial T} = -\frac{\bar{\tilde{H}}_i}{T^2} \qquad [6.10]$$

$\bar{\tilde{H}}_i$ is the *partial molar electrochemical enthalpy* of the component i.

6.1.3.4. *Variations of the electrochemical potential with changing composition*

The properties of symmetry of the characteristic matrix also give us:

$$\frac{\partial\tilde{\mu}_i}{\partial n_j} = \frac{\partial\tilde{\mu}_j}{\partial n_i} \qquad [6.11]$$

6.1.3.5. *Total differential of the electrochemical potential*

By combining the relations obtained by applying expression [6.6] to the pressure and temperature with relation [6.11], we find:

$$\mathrm{d}\tilde{\mu}_i = -T\,\mathrm{d}\tilde{S} + P\,\mathrm{d}V - \Phi\,\mathrm{d}q + \sum_{j=1}^{N}\frac{\partial\tilde{\mu}_i}{\partial n_j}\mathrm{d}n_j \qquad [6.12]$$

6.1.4. *Gibbs–Duhem relation for electrochemical systems*

Because the electrochemical potential is a partial molar value, we can write that it represents the electrochemical Gibbs molar energy, and:

$$\tilde{G} = \sum_{i=1}^{N}\tilde{\mu}_i n_i \qquad [6.13]$$

From this, we deduce the Gibbs–Duhem equation in its electrochemical form:

$$\sum_{i=1}^{N} n_i \,\mathrm{d}\tilde{\mu}_i - \tilde{S}\,\mathrm{d}T - q\,\mathrm{d}\Phi + V\,\mathrm{d}P = 0 \qquad [6.14]$$

and its simplified form where the physical intensive variables are kept constant:

$$\sum_{i=1}^{N} n_i \,\mathrm{d}\tilde{\mu}_i = 0 \qquad [6.15]$$

In light of relation [6.12], this last relation gives us:

$$\sum_{j=1}^{N} \frac{\partial \tilde{\mu}_i}{\partial n_j} \mathrm{d}n_j = 0 \qquad [6.16]$$

6.1.5. *Chemical system associated with an electrochemical system*

If we consider an electrochemical system formed of one phase, whose la description requires the use of the couples of variables T, S; P, T; and Φ, q, the chemical system associated with that electrochemical system is the identical system in which the variables Φ and q are not taken into account, or – which amounts to the same thing – in which one of the two variables in the couple q, Φ (whichever is chosen as the variable for the problem) has the value of zero.

The generalized Gibbs energy of the system is the electrochemical Gibbs energy $\tilde{G}$ defined by relation [6.5].

Consider the function of state defined by:

$$U_{chem} = \tilde{U} - q\Phi \qquad [6.17]$$

U_{chem} is the internal energy of the chemical system associated with our electrochemical system.

By differentiating at null potential, we find:

$$\mathrm{d}U_{chim} = T\,\mathrm{d}S - P\,\mathrm{d}V + \sum \mu_i\,\mathrm{d}n_i \qquad [6.18]$$

μ_i is the chemical potential of i in the chemical system associated with our electrochemical system. We see that we can write:

$$\tilde{\mu}_i = \mu_i + q\Phi \qquad [6.19]$$

However, if $\mathfrak{F}$ denotes the Faraday and z_i the electrovalence of the component i, the charge is:

$$q = \sum n_i z_i \mathfrak{F} \qquad [6.20]$$

When we substitute back into relation [6.19], the electrochemical potential becomes:

$$\tilde{\mu}_i = \mu_i + \mathfrak{F}\Phi \sum_i n_i z_i \qquad [6.21]$$

In view of the remark in section 6.1.1, it is not certain that the chemical potential of a species given by relation [6.21] will be the chemical potential of that same species in the absence of field, because of the entropy term, which does not necessarily have to be the same in the two cases. Thus, it is preferable, as stipulated in 1961 by the *Comité International de Thermodynamique Electrochimique* (which, in 1971, become the *Société Internationale d'Electrochimie*), to define the chemical values associated with the electrochemical system by relations of the same type as relation [6.17].

6.1.6. *General conditions of an equilibrium of an electrochemical system*

Consider an electrochemical reaction between electrons and ions. We write it in the more general form in accordance with:

$$\sum_i \nu_i \mathrm{A}_i + \nu_e e = 0 \qquad [6R.1]$$

Thus, we distinguish the electrons from the other components A_i (ions, molecules and atoms). The coefficients ν_i are positive or negative, and the coefficient of the electrons ν_e is taken as positive, meaning that the reactions are always written in the direction of reduction.

The electrochemical affinity of the reaction will be defined by a relation similar to that used in the chemical systems, which means that we can write the relation:

$$\tilde{\mathscr{A}}_i = -\sum_i \nu_i \tilde{\mu}_i - \nu_e \tilde{\mu}_e \qquad [6.22]$$

The equilibrium condition of the electrochemical system should render a zero value of the electrochemical affinity:

$$\tilde{\mathscr{A}}_i = 0 \qquad [6.23]$$

In light of relation [6.22], the equilibrium condition is written:

$$\sum_i \nu_i \tilde{\mu}_i + \nu_e \tilde{\mu}_e = 0 \qquad [6.24]$$

We shall now examine one initial application of this condition, considering the system to be single-phased, meaning that the reaction takes place within a phase which contains both ions and electrons and placed at an electrical potential Φ. Taking account of the expression [6.21] of an electrochemical potential $\tilde{\mu}_i$ as a function of the chemical potential μ_i, the equilibrium condition becomes:

$$\sum_i \nu_i \mu_i + \nu_e \mu_e + \mathfrak{F}\Phi\left(\sum_i \nu_i z_i - \nu_e\right) = 0 \qquad [6.25]$$

The electrical neutrality of the system means that we can write:

$$\sum_i \nu_i z_i - \nu_e = 0 \qquad [6.26]$$

The equilibrium relation is then simplified to:

$$\sum_i \nu_i \mu_i + \nu_e \mu_e = 0 \qquad [6.27]$$

We can see that this condition is independent of the electrical potential of the phase, and is written in the same way as the condition of a classic chemical equilibrium, considering the electrons to be one of the components of the system.

By expressing the chemical potentials as a function of the activities of the species, we obtain the law of mass action in the form:

$$\prod_i |\mathrm{A}_i|^{\nu_i} |e| = K(T) \qquad [6.28]$$

We shall now give a second application of [6.24], in the case of the equilibrium of two metal phases in contact with one another. In a system such as this, only the electrons are able to pass from one metal phase (1) into the other metal phase (2). Thus, the equilibrium condition will be written:

$$\mu_e^{(1)} - \mathfrak{F}\Phi_1 = \mu_e^{(2)} - \mathfrak{F}\Phi_2 \qquad [6.29]$$

From this, we deduce:

$$\Phi_1 - \Phi_2 = \frac{\mu_e^{(1)} - \mu_e^{(2)}}{\mathfrak{F}} = \text{constant} \qquad [6.30]$$

Hence, there is a so-called *contact potential difference* between two metal phases in contact. This potential difference is around 1 volt.

6.2. The electrode

In this next part, we turn our attention to the most common type of electrochemical system: the electrode.

6.2.1. *Definition and reaction of the electrode*

Electrode denotes the diphasic system comprising an electronic conductor (made of metal or a semi-conductor) (phase 1 with potential Φ_1) in contact with an ionic conductor, called the electrolyte (phase 2 with potential Φ_2). That ionic conductor is a medium which contains ions with the ability to move. It may be solid, liquid or gaseous. The medium may be a pure phase (a molten salt, an ionic solid) or a solution (aqueous liquid solution, solid solution) in which certain components are ionized.

We suppose that our system is host to reaction [6R.1]. The electrons exist only in phase 1. On the other hand, out of the components A_i, the ions are only present in phase 2. The neutral molecules may be present in both phases.

As the system is supposed to be closed (insulated electrode), the reaction will entail the transfer of electrons from phase 2 to phase 1, accompanied by corresponding variations in the concentrations of the components A_i.

We use $A_i^{(1)}$ to represent the substances present in phase 1 (they are neutral) and $A_i^{(2)}$ to represent those which are present in phase 2 (some of these are neutral, and others ionized).

An electrode is denoted by a symbol system showing the two phases and the inter-phase surface such as, e.g. the electrode made of a sheet of silver, immersed in an aqueous solution containing Ag+ ions, which we write as:

$$Ag^+ | Ag$$

Sometimes such an electrode is called a *silver electrode*.

In the case of an electrode, the variable pressure will be the pressure in the volume in each of the phases situated on both sides of the interface.

6.2.2. *Equilibrium of an insulated metal electrode – electrode absolute voltage*

The equilibrium condition [6.27], applied to equilibrium [6R.1], in view of our notations and explicitly stating the different terms, is written:

$$\sum_i \nu_i \mu_i^{(1)} + \nu_e \mu_e^{(1)} - \mathfrak{F}\Phi_1 \nu_e + \sum_j \nu_j \mu_j^{(2)} + \mathfrak{F}\Phi_2 \sum_j \nu_j z_j = 0 \qquad [6.31]$$

When we take account of the electrical neutrality linked to the reaction, and combine the terms in $\mu_i^{(1)}$ and $\mu_j^{(2)}$, this relation becomes:

$$\sum_i \nu_i \mu_i + \nu_e \mu_e^{(1)} - \mathfrak{F} \nu_e \left(\Phi_1 - \Phi_2 \right) = 0 \qquad [6.32]$$

The potential difference $\Phi_1 - \Phi_2$ is called the *electrode absolute voltage* e_{abs}, and we have:

$$e_{\text{abs}} = \frac{1}{\nu_e \mathfrak{F}} \sum_i \nu_i \mu_i + \frac{\mu_e^{(1)}}{\mathfrak{F}} \qquad [6.33]$$

The term $\mu_e^{(1)}$, which is the chemical potential of the electrons in the metal, is unknown but very large, and thus can be considered a constant characteristic of the metal.

NOTE.– For electrodes with semi-conductors, the effective electron concentration must be taken into account to determine the chemical potential of the electrons, because that concentration can no longer be considered a constant (see Chapter 3 in Volume 3 of this set [SOU 15c]).

6.2.3. Voltage relative to a metal electrode – Nernst's relation

The absolute electrode voltage defined above, which is an internal potential difference, cannot be measured directly. In addition, it is not possible to calculate it, because we are unable to precisely determine the chemical potentials of the components A_i. The only known method is to determine relative chemical potentials, by means of a reference electrode. All we can do, then, is to compare the absolute voltage of one electrode to that of another electrode chosen of the reference electrode, and thus define the *relative voltage*.

We shall come back later on (in sections 6.3.2.2 and 6.3.3) to the two main reference electrodes used: the calomel electrode and the hydrogen electrode, which constitutes the primary reference.

Thus, we associate our electrode with a standard hydrogen electrode (gas pressure of 1 bar), creating the following electrochemical chain:

A	|	$\sum_i A_i$	||	H^+	|	$Pt\|H_2\|$	|	A
Phase 1		Phase 2		Phase 2'		Phase 1'		Phase 1"

The presence of two vertical lines represents the presence of a membrane which allows ions to pass through, but does not allow the two phases, which are usually liquids, to mix into one.

Note that in the electrochemical chain thus defined, the two extremities of the cell are formed of the *same metal element*, because the voltmeter can only measure the potential difference between two points with the same conductor. In fact, therefore, our chain measures the potential difference $\Phi_1 - \Phi_{1''}$, and it is that difference which is called the *relative voltage of the electrode* and is denoted by *e*. Thus:

$$e = \Phi_1 - \Phi_{1''} \qquad [6.34]$$

That relative voltage can, in fact, be considered to be the sum of four interphase voltages, and be written in the form:

$$e = (\Phi_1 - \Phi_2) + (\Phi_2 - \Phi_{2'}) + (\Phi_{2'} - \Phi_{1'}) + (\Phi_{1'} - \Phi_{1''}) \qquad [6.35]$$

The difference $\Phi_2 - \Phi_{2'}$ is called the *liquid junction potential difference* (see section 6.4) and can be rendered negligible using a salt bridge constituted by two very similar ions such as potassium chloride, rubidium bromide or cesium iodide. The differences $\Phi_1 - \Phi_2$ and $\Phi_{1'} - \Phi_{2'}$ are the absolute voltages of the electrode under study and of the reference hydrogen electrode.

We shall see (section 6.3.3) that this latter electrode is the site of the electrochemical reaction:

$$H^+ - \frac{1}{2}H_2 + e = 0 \qquad [6R.2]$$

The difference $\Phi_{1'} - \Phi_{1''}$ is a contact potential difference which can be written in the form:

$$\Phi_{1'} - \Phi_{1''} = \frac{\mu_e^{(1')} - \mu_e^{(1'')}}{\mathfrak{F}} = \text{Constant} \qquad [6.36]$$

In light of relations [6.33] and [6.36], we obtain:

$$e = \frac{\mu_e^{(1)}}{\mathfrak{F}} + \frac{1}{\nu_e \mathfrak{F}} \sum_{i-} \nu_i \mu_i - \frac{\mu_e^{(1')}}{\mathfrak{F}} - \frac{1}{\mathfrak{F}}\left(\mu_{\mathrm{H}^+}^0 - \frac{1}{2}\mu_{\mathrm{H}_2}^0 \right) + \frac{\mu_e^{(1')}}{\mathfrak{F}} - \frac{\mu_e^{(1'')}}{\mathfrak{F}} \qquad [6.37]$$

The sum $\sum_{i-} \nu_i \mu_i$ takes account of all of the species with the exception of the electrons.

In standard conditions at equilibrium, we have:

$$\mu_{\mathrm{H}^+}^0 - \frac{1}{2}\mu_{\mathrm{H}_2}^0 = 0 \qquad [6.38]$$

Furthermore, the chemical potential of the electrons is the same in both phases 1 and 1'', which are identical, formed of the same metal. Thus, we finally obtain:

$$e = \frac{1}{\nu_e \mathfrak{F}} \sum_{i-} \nu_i \mu_i \qquad [6.39]$$

If we write the chemical potentials as a function of the activities (or of the fugacities in the case of gases), we find Nernst's classic formula:

$$e = \frac{1}{\nu_e \mathfrak{F}} \sum_{i-} \nu_i \mu_i^0 + \frac{\mathrm{R}T}{\nu_e \mathfrak{F}} \ln \prod |\mathrm{A}_i|^{\nu_i} \qquad [6.40]$$

Here, e would simply be called the *electrode voltage*. We extract the quantity:

$$e^0 = \frac{1}{\nu_e \mathfrak{F}} \sum_{i-} \nu_i \mu_i^0 \qquad [6.41]$$

This voltage is called the *normal voltage* or *standard potential of the electrode*, and Nernst's relation [6.40] takes its conventional form:

$$e = e^0 + \frac{\mathrm{R}T}{\nu_e \mathfrak{F}} \ln \prod |\mathrm{A}_i|^{\nu_i} \qquad [6.42]$$

If the conditions of concentration and total pressure allow for it, the activities and fugacities can be replaced respectively by the concentrations and the partial pressures depending on the nature of the phase. If we choose the standard pressure of 1 bar, only the solutes come into play, and Nernst's relation then becomes:

$$e = e_0 + \frac{\mathrm{R}T}{\nu_e \mathfrak{F}} \ln \prod [\mathrm{A}_i]^{\nu_i} \qquad [6.43]$$

This is the primitive form of Nernst's relation.

6.2.4. Chemical and electrochemical Gibbs energy of the electrode reaction

In view of the relation between the Gibbs energy ΔG_{red} associated with the reaction [6R.1] and the chemical potentials of the components, relation [6.39] immediately gives us:

$$e = -\frac{\Delta G_{\mathrm{red}}}{\nu_{\mathrm{e}} \mathfrak{F}} \qquad [6.44]$$

Similarly, in standard conditions, relation [6.41] yields:

$$e^0 = -\frac{\Delta G^0_{\mathrm{red}}}{\nu_{\mathrm{e}} \mathfrak{F}} \qquad [6.45]$$

Relations [6.44] and [6.45] constitute a second form of Nernst's relation.

Consider the sum:

$$\Delta G_{\mathrm{red}} + \nu_{\mathrm{e}} \mathfrak{F} e = \Delta \tilde{G}_{\mathrm{red}} \qquad [6.46]$$

That sum, which takes the value of 0 at equilibrium, constitutes an electrochemical enthalpy associated with the reduction reaction [6R.1]. However, be aware that this electrochemical Gibbs energy is a relative electrochemical Gibbs energy, because it is defined on the basis of relative chemical potentials and varies in the same direction as the absolute electrochemical Gibbs energy, defined in section 6.1.2. In practice, though, it is the relative Gibbs energy which is known as the electrochemical Gibbs energy associated with the electrode.

Based on that electrochemical Gibbs energy, we define the *electrochemical affinity* of the reduction by:

$$\tilde{\mathscr{A}}_{red} = -\Delta \tilde{G}_{\text{red}} \qquad [6.47]$$

Here, again, we are dealing with a relative electrochemical affinity, but it is commonly referred to simply as the electrochemical affinity.

6.2.5. *Influence of pH on the electrode voltage*

If the components involved in the reaction [6R.1] of reduction of the electrode include protons, then Nernst's relation [6.40] will involve the logarithm of the concentration of protons, and thus the electrode voltage will be contingent upon the pH.

We can illustrate this by choosing an electrode which works by the following reaction of reduction of the permanganate ion in the system Mn^{2+}/MnO^{-}_{4}:

$$\text{MnO}_4^- + 8\text{H}^+ + 5\text{e} = \text{Mn}^{2+} + 4\text{H}_2\text{O} \qquad [6\text{R}.3]$$

The application of relation [6.40] gives us:

$$e_{(3)} = e^0_{(3)} + \frac{RT}{5\mathscr{F}} \ln \frac{\left|\text{MnO}_4^-\right|\left|\text{H}^+\right|^8}{\left|\text{Mn}^{2+}\right|} \quad (\text{where } e^0_{(3)} = 1.51 \text{ V}) \qquad [6.48]$$

and by explicitly writing the pH, we find:

$$e_{(3)} = e^0_{(3)} - \frac{8}{5} 0.06\text{pH} + \frac{RT}{5\mathscr{F}} \ln \frac{\left|\text{MnO}_4^-\right|}{\left|\text{Mn}^{2+}\right|} \qquad [6.49]$$

If the activities of both the oxidized and reduced forms are equal, the voltage takes the value e_0, which is called the *equi-activity voltage*, and is given by:

$$e_{0(3)} = e^0_{(3)} - \frac{8}{5}0.06\text{pH} \qquad [6.50]$$

In the coordinates $e_{0(3)} = f(\text{pH})$, we obtain a straight line whose slope is $-\frac{8}{5}0.06 = -0.096$, which constitutes the voltage/pH curve for the system Mn^{2+}/MnO^{4-} (Figure 6.1).

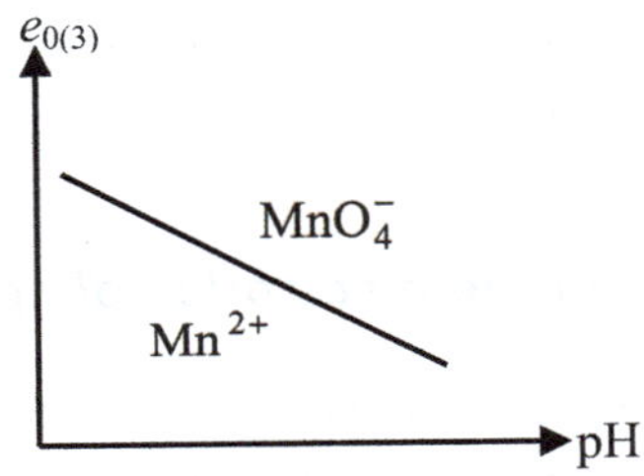

Figure 6.1. *Voltage/pH curve for a redox couple such as Mn^{2+}/MnO^{4-}*

That line splits the space into two regions. Each point in the space corresponds to a specific value of the ratio $|MnO_4^-|/|Mn^{2+}|$, which can be calculated using Nernst's relation written in the form:

$$e_{(3)} = e_{0(3)} + \frac{RT}{5\mathfrak{F}} \ln \frac{|MnO_4^-|}{|Mn^{2+}|} \qquad [6.51]$$

We see that:

– if $e_{(3)} > e_{0(3)}$ then $|MnO_4^-| > |Mn^{2+}|$

– if $e_{(3)} < e_{0(3)}$ then $|MnO_4^-| < |Mn^{2+}|$

The half-space above the straight line corresponds to the predominance of the oxidized form $\left|MnO_4^-\right|$, whilst the half-space below the straight line corresponds to the predominance of the reduced form $\left|Mn^{2+}\right|$.

In fact, as soon as the deviation from the straight line is a few hundred volts or more, the predominant form is encountered in massive quantities. For this reason, we often speak of the two half-spaces, which are simply domains of predominance, as *domains of stability* of one form or the other.

6.2.6. *Influence of the solvent and of the dissolved species on the electrode voltage*

Water, or one of its ions, may react with a redox couple, and thereby modify the voltages. We shall take the example of an electrode Cu^{2+}/Cu, whose reduction reaction is:

$$Cu^{2+} + 2e = Cu(s) \qquad [6R.4]$$

According to Nernst, its voltage is written;

$$e_{Cu^{2+}/Cu} = e^0_{Cu^{2+}/Cu} - \frac{RT}{2\mathcal{F}} \ln\left|Cu^{2+}\right| \qquad [6.52]$$

Yet in a weakly acidic or weakly basic medium, water reacts with copper ions, as follows:

$$Cu^{2+} + 2H_2O = Cu(OH)_2(s) + 2H^+ \qquad [6R.5]$$

The constant for such an equilibrium is expressed in the form;

$$K = \frac{\left|H^+\right|^2}{\left|Cu^{2+}\right|} \qquad [6.53]$$

By combining relations [6.52] and [6.53], we find:

$$e_{Cu(OH)_2/Cu} = e^0_{Cu^{2+}/Cu} - \frac{RT}{2\mathcal{F}} \ln K - \frac{RT}{\mathcal{F}} \ln\left|H^+\right| \qquad [6.54]$$

In addition, if we consider the electrode working by way of the reaction:

$$\mathrm{Cu(OH)_2}(s) + 2\mathrm{H}^+ + 2\mathrm{e} = \mathrm{Cu}(s) + 2\mathrm{H_2O} \qquad [6R.6]$$

its potential is given by:

$$e_{\mathrm{Cu(OH)_2/Cu}} = e^0_{\mathrm{Cu(OH)_2/Cu}} - \frac{\mathrm{R}T}{\mathcal{F}} \ln\left|\mathrm{H}^+\right| \qquad [6.55]$$

By combining expressions [6.54] and [6.55], we obtain:

$$e^0_{\mathrm{Cu(OH)_2/Cu}} = e^0_{\mathrm{Cu^{2+}/Cu}} + \frac{\mathrm{R}T}{2\mathcal{F}} \ln K \qquad [6.56]$$

Thus, the normal potentials of the two couples $Cu(OH)_2/Cu$ and Cu^{2+}/Cu are not independent.

Similarly, an element of a redox couple may react with an ion present in the solution. We shall look at the case of the couple Ag^+/Ag in the presence of a bromide ion. The reaction of the electrode is written:

$$\mathrm{Ag}^+ + \mathrm{e} = \mathrm{Ag} \qquad [6R.7]$$

The reaction between the silver ions and bromide ions leads to the precipitation of solid silver bromide, in a pure phase, thus:

$$\mathrm{Ag}^+ + \mathrm{B}r^- = \mathrm{AgBr}(s) \qquad [6R.8]$$

Such an equilibrium is characterized by the solubility product K_s of the silver bromide.

Employing the same reasoning as in the previous example, we find the relation between the normal potentials of the two couples AgBr/Ag and Ag^+/Ag, which are therefore not mutually independent:

$$e^0_{\mathrm{AgBr/Ag}} = e^0_{\mathrm{Ag^+/Ag}} + \frac{\mathrm{R}T}{\mathcal{F}} \ln K_s \qquad [6.57]$$

Using this expression, we are able either to calculate one normal potential if we know the other and the solubility product, or to determine the solubility product on the basis of the normal potentials of the two couples.

6.2.7. *Influence of temperature on the normal potentials*

Let us examine the variations in the absolute normal potentials with changing temperature. By applying the classic relation of the variation in Gibbs energy with temperature, for reaction [6R.1], we find the relation:

$$\frac{\mathrm{d}\Delta G^0}{\mathrm{d}T} = -\Delta S^0 \qquad [6.58]$$

By applying expression [6.45] to the absolute normal potentials, and combining it with expression [6.58], we deduce the relation which gives the variations in the absolute normal potential of an electrode associated with the reaction [6R.1]:

$$\frac{\mathrm{d}e^0_{\mathrm{abs}}}{\mathrm{d}T} = -\frac{\Delta S^0}{\nu_e \mathcal{F}} \qquad [6.59]$$

Remember that for the hydrogen electrode reference, we set $e^0_{\mathrm{H}} = 0$ at the temperature of 25°C. That potential obviously varies with temperature, obeying a law similar to [6.59], but to quantify that variation, we need to enact the following chain:

$$\begin{matrix} \{\mathrm{H_2}\} \mid \langle\mathrm{Pt}\rangle \mid (\mathrm{H^+}) \mid \langle\mathrm{Pt}\rangle \mid \{\mathrm{H_2}\} \\ (T_1) \qquad\qquad\qquad (T_2) \end{matrix}$$

The two electrodes are placed at two different temperatures. Then, though, an experimental difficulty emerges because, in view of the existence of the temperature gradient, the phenomenon of thermo-diffusion occurs in the electrolyte, preventing electrochemical equilibrium from being reached.

Faced with this difficulty, we make the decision to extend the previous convention, setting that $e^0_{\mathrm{H}} = 0$ at *all temperatures*.

Thus, the relative potentials of electrodes will vary with temperature by way of a law in the form:

$$\frac{\mathrm{d}e^0}{\mathrm{d}T} = -\frac{\Delta_1 S^0_{\mathrm{rel}}}{\nu_e \mathfrak{F}} \quad [6.60]$$

In this law, the entropy variation is obviously the relative entropy variation – i.e. the variation in entropy of reaction [6R.1] corrected by the corresponding term for reaction [6R.2]. Thus, we have:

$$\frac{\mathrm{d}e^0}{\mathrm{d}T} = -\frac{\Delta_1 S^0 - \nu_e \Delta_2 S^0}{\nu_e \mathfrak{F}} \quad [6.61]$$

For example, with a calomel electrode in the presence of a saturated potassium chloride solution (see section 6.3.2.2), it can be shown that the differential of $\mathrm{d}e^0/\mathrm{d}T$ has the value of -76×10^{-5} at 25°C.

The tables give the temperature coefficients $\mathrm{d}e^0/\mathrm{d}T$ for a number of simple electrodes; these coefficients are obtained on the basis of isothermal cells using the overall convention adopted above.

6.3. The different types of electrodes

We distinguish between three categories of electrodes, depending on the nature of the electrochemical reaction which makes them work:

– so-called "redox" electrodes;

– metal electrodes;

– gas electrodes.

6.3.1. *Redox electrodes*

In this type of electrode, the metal – often platinum – serves only to conduct the electrons.

6.3.1.1. *Definition*

Clearly, all electrodes work on the basis of a redox reaction, but the name of *redox electrode* is reserved for electrodes composed of a metal wire immersed in a solution containing different components; only those components actually participate in the electrochemical reaction; we say that they are electro-active. The metal is not electro-active, because it is not involved in the reaction by which the electrode functions. As an example, we can look at an electrode which has the structure $Pt/Fe^{2+}/Fe^{3+}$, whose functional reaction is:

$$Fe^{3+} + e = Fe^{2+} \quad [6R.9]$$

6.3.1.2. *Redox buffer effect*

Let us take the example of the couple Fe^{2+}/Fe^{3+} ($e_0 = 0.77V$), and suppose that the total concentration of iron is equal to 1mole/l. We simplify our thought process by treating the concentrations and activities as being the same. The respective concentrations of Fe^{2+} and Fe^{3+} would be x and $1-x$. The platinum electrode immersed in that solution takes the voltage:

$$e_{(9)} = 0.77 + \frac{RT}{\mathfrak{F}} \ln \frac{x}{(1-x)} \quad [6.62]$$

Figure 6.2 illustrates the variations in that voltage with x. It shows an inflection point which corresponds to $x = ½$, which is iso-composition in terms of ions. At that point, as the tangent is fairly flat, the voltage varies very little through the addition of an oxidant or reductant. This mirrors the properties of acid/base buffer solutions; here, though, we are dealing with a *redox buffer solution*. The buffer power is defined by the differential de/dx, whose value is:

$$\frac{d e_{(9)}}{d x} = \frac{RT}{\mathfrak{F}} x(1-x) \quad [6.63]$$

6.3.1.3. *Limitation of voltages due to the solvent*

Theoretically, according to relation [6.62], if the oxidant or reductant are alone in the solution, the electrode must take a voltage of between $-\infty$ and $+\infty$. In fact, there is one initial phenomenon that occurs which limits that voltage – this is an electro-active role of the solvent.

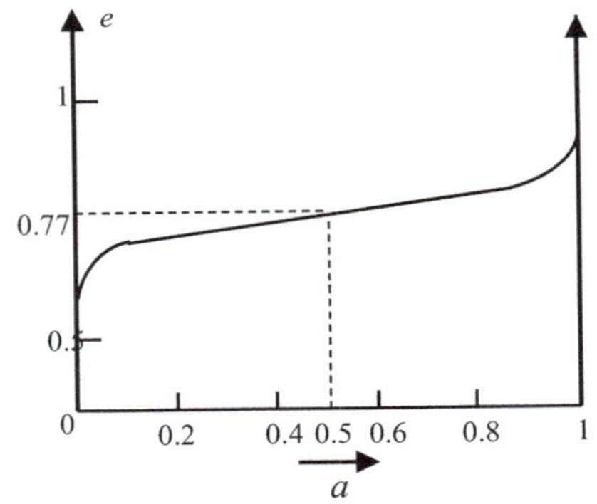

Figure 6.2. *Variations of the redox potential of the couple Fe2+/Fe3+ as a function of the activities*

Indeed, no solvent is ideal, being both ionizing and non-electro-active. If we take a solution of the oxidant alone, it will be reduced, and a small amount of reductant will appear until the voltages of the system in question and of the solvent/oxidized solvent system are equalized. In the case of water, for example, an electrochemical reaction of oxidation of the solvent is written:

$$H_2O = 1/2O_2 + 2H^+ + 2e \qquad [6R.10]$$

At atmospheric pressure, and at a pH of 0, the voltage of that system is 1.23 V. In the presence of a salt of Fe^{3+}, we have the equilibrium [6R.10] which is established, and thus:

$$0.77 + \frac{RT}{\mathcal{F}} \ln \frac{x}{1-x} = 1.23 \qquad [6.64]$$

Hence, for a concentration:

$$x = \frac{10^{7.66}}{1+10^{7.66}} \cong 1 \qquad [6.65]$$

Now, if the solution contains only a salt of Fe^{2+}, the reduction of water takes the form of reaction [6R.2].

However, conventionally, this system has a voltage of 0 at a pH of 0, and at a hydrogen pressure of 1 bar. The equilibrium of the voltages then tells us that:

$$0.77+\frac{RT}{\mathcal{F}}\ln\frac{x}{1-x}=0 \qquad [6.66]$$

Thus:

$$x=\frac{10^{-12.8}}{1+10^{-12.8}}\cong 0 \qquad [6.67]$$

Hence, the voltage interval (in our case, 0–1.23 V) in which the solvent has no electro-active role constitutes the *domain of electro-activity corresponding to that solvent*. This is the domain in which we can measure the potential of the redox electrode.

6.3.1.4. *Limitation of voltages due to dilution*

Experimentally, we find that if one of the oxidized or reduced forms is greatly diluted, it ceases to be electro-active. In these conditions, the electrode assumes a voltage which depends only on the activity of the non-diluted form by way of a law similar to Nernst's equation, but with empirical constants which appear to be kinetic in origin. We then have:

– on the side of the oxidant ($x = 1$):

$$e=A+\frac{RT}{\mathcal{F}}\ln x\cong A-\frac{RT}{\mathcal{F}}(1-x) \qquad [6.68]$$

– on the side of the reductant ($x = 0$):

$$e=B-\frac{RT}{\mathcal{F}}\ln(1-x)\cong B+\frac{RT}{\mathcal{F}}x \qquad [6.69]$$

These voltages tend respectively toward the values A and B.

We can see that this limitation due to dilution occurs long before the limitation due to the solvent, seen in section 6.3.1.3. For example, in the case of the couple Fe^{2+}/Fe^{3+}, we note in practice that for concentrations of 10^{-6} mol/l, the Fe^{2+} or Fe^{3+} ions become inactive. Using relations [6.68] and

[6.69] (with $A = B = 0.77$ V), we can easily calculate that this corresponds to an interval of electro-activity of (0.14–1.13 V).

6.3.1.5. *Redox electrode with the involvement of water ions: the quinhydrone electrode*

The exchange of electrons between the two redox forms can, of course, also involve the ions of water, as we saw in the case of the couple Mn^{2+}/MnO_4^- (see section 6.2.5).

Such is the case in the quinhydrone electrode, which is therefore used for measuring pH.

Quinhydrone is a crystallized equimolecular mixture of benzoquinone and hydroquinone.

That electrode (see Figure 6.3) operates by the reaction:

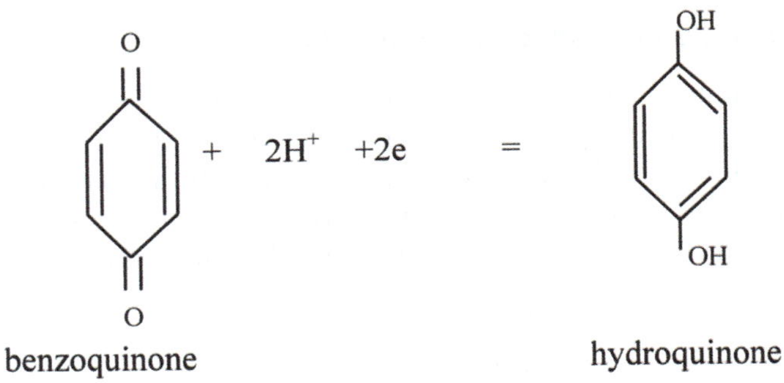

[6R.11]

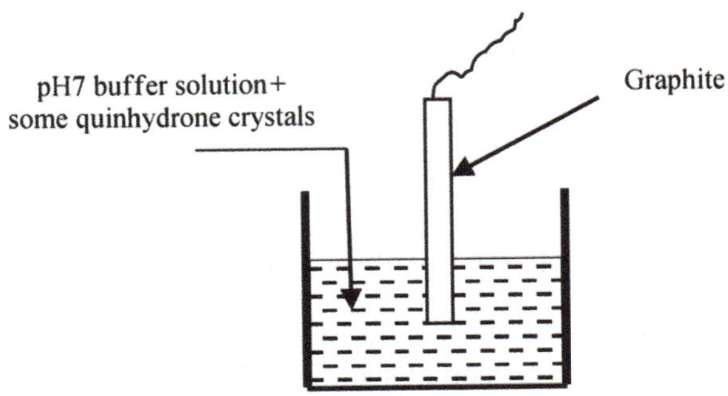

Figure 6.3. *Diagram of the quinhydrone electrode*

More generally speaking, an electrode using the ions of water operates by the reaction:

$$\mathrm{Ox} + q\mathrm{H}^+ + \nu_e \mathrm{e} = \mathrm{Red} \qquad [6R.12]$$

We find the relation:

$$e = e^0 + \frac{\mathrm{R}T}{\nu_e \mathcal{F}} \ln \frac{[\mathrm{OX}]\left[\mathrm{H}^+\right]^q}{[\mathrm{Red}]} \qquad [6.70]$$

and in the case of quinone:

$$e = e^0 + \frac{\mathrm{R}T}{2\mathcal{F}} \ln \frac{[\mathrm{Quinone}]\left[\mathrm{H}^+\right]^2}{[\mathrm{Hydroquinone}]} \qquad [6.71]$$

That electrode is also used as a reference electrode. Its normal voltage is $e^0 = 0.699$ V.

In a non-aqueous medium, we substitute quinhydrone with the couple tetrachlorohydroquinone (or chloranile)/hydrochloranile, which works better as a pH electrode.

6.3.2. *Metal electrodes*

Metal electrodes are classified into *different species* depending on the number of solid phases involved in the electrochemical reaction.

6.3.2.1. *First-species metal electrodes*

In a first-species metal electrode, there is only one solid phase: the metal. The simplest such electrodes are made up of the metal at equilibrium with a solution containing an ion of the same element – e.g. the couple Fe/Fe^{2+}.

Such an electrode is easy to create, because the metal does not attack the solvent. In water, for example, the electrode's voltage must be in the corresponding domain of electro-activity – i.e. at a pH of 0, between 0 V and 1.23 V.

If the metal attacks the water, we can still, in certain cases, indirectly obtain the experimental measurement of the voltage.

In certain cases, the simple ion of the overly-acidic metal does not exist in solution, and it is only possible to create a first-species electrode in the presence of a complexant of that ion. For instance, with titanium, we create a first-species electrode:

$$\mathrm{Ti} \mid [\mathrm{TiF_6}]^{2-}$$

That electrode is based on the following electrochemical reaction:

$$\mathrm{Ti} + \mathrm{F}^- = [\mathrm{TiF_6}]^{2-} + 4\mathrm{e} \qquad [6R.13]$$

All of the normal voltages of simple first-species electrodes are quoted in tables, known as Nernst's tables.

6.3.2.2. *Second-species metal electrodes*

A second-species metal electrode is one in which the metal is at equilibrium with a second solid phase containing the same element as the metal and a solution having an ion of an element of the second solid phase.

The best-known example is the calomel electrode (calomel is mercury(I) chloride: Hg_2Cl_2):

$$(\mathrm{Hg}) \mid <\mathrm{Hg_2Cl_2}> \mid (\mathrm{Cl}^-)$$

This electrode serves as a secondary reference for measuring electrode potentials, because second-species metal electrodes are indeed reversible and have low polarizability. Figure 6.4 illustrates such an electrode.

The calomel electrode works on the basis of the following electrochemical reaction:

$$2(\mathrm{Hg}) + 2(\mathrm{Cl}^-) = \langle \mathrm{Hg_2Cl_2} \rangle + 2\mathrm{e} \qquad [6R.14]$$

Of course, the chloride solution is saturated with calomel, which corresponds to the solubility product $K_S = |Hg^{2+}||Cl^{-2}| = 2\times10^{-18}$ mol/l at 25°C.

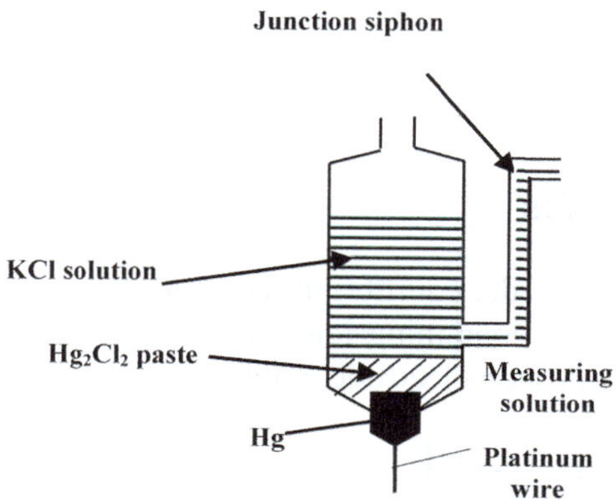

Figure 6.4. *Diagram of the calomel electrode*

Its voltage then obeys the expression:

$$e = e^0 + \frac{RT}{2\mathfrak{F}}\ln|Hg^{2+}| = e^0 + \frac{RT}{2\mathfrak{F}}\ln K_S - \frac{RT}{\mathfrak{F}}\ln|Cl^-| \qquad [6.72]$$

Thus, if we use the numerical values:

$$e = 0.25 - 0.06\ln|Cl^-| \qquad [6.73]$$

Thus, we can see that the calomel electrode is an indicator electrode for the chlorine ion, in the sense that by measuring the voltage, we are able to determine the solution's chlorine activity.

This property is generally applicable to all second-species metal electrodes, which act as indicator electrodes for the anion of the solid salt.

More generally, we can define n^{th}-species metal electrodes. An electrode such as this would be constituted by a metal at equilibrium with n–1 solid phases, the first of which contains the cation corresponding to the metal, and pairwise, those solid phases have, alternately, an anion and a cation in common. If the number n is odd, those electrodes are indicators for the last cation in the chain; if n is even, they are indicators for the last anion in the chain. The following example represents a 5th-species electrode:

$$\mathrm{Zn} \mid \langle \mathrm{ZnC_2O_4} \rangle \mid \langle \mathrm{Ag_2C_2O_4} \rangle \mid \langle \mathrm{AgCl} \rangle \langle \mathrm{TlCl} \rangle \mid \mathrm{Tl^+}$$

Such an electrode can be used to measure the activity of the Tl^+ ions.

6.3.3. *Gas electrodes*

In this type of electrode, equilibrium is created between a gas at a certain pressure and a solution containing ions of the same element as the gas. The best known is the hydrogen electrode, which operates by reaction [6R.2]:

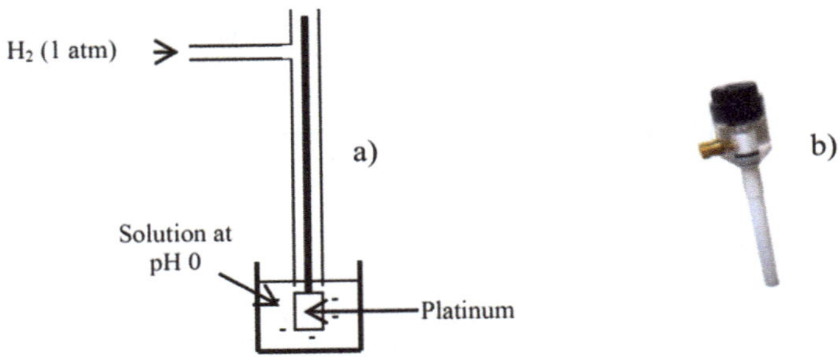

Figure 6.5. *Reference hydrogen electrode: a) Diagram showing the principle; b) commercially-available electrode*

Figure 6.5(a) illustrates the principle of a normal hydrogen electrode (NHE): hydrogen, at a pressure of 1 bar, bubbles through a solution of an acid at pH zero, into which a platinum electrode is immersed, chose purpose is to provide the external electrical contact. Figure 6.5(b) shows a photo of a commercial hydrogen electrode.

By applying Nernst's relation, we find the voltage of a hydrogen electrode to be:

$$e = e^0_{(\mathrm{H})} + \frac{\mathrm{R}T}{\mathcal{F}} \ln \frac{|\mathrm{H}^+|}{P^{1/2}_{\mathrm{H_2}}} \qquad [6.74]$$

As the normal voltage is taken as a reference ($e^0_{(H)} = 0$), the practical value of the voltage of a hydrogen electrode is as follows, when expressed through decimal logarithms:

$$e = -0.06\text{pH} - 0.03\log P_{H_2} \quad [6.75]$$

It is also common to use oxygen electrodes. The electrolyte is then a solid metal oxide, which is an ionic conductor by oxygen ions such as yttrium-stabilized zirconia. This electrode works through the reaction:

$$1/2O_2 + 2e + 2H^+ = H_2O \quad [6R.15]$$

Its voltage is given by:

$$e = 1.23 - 0.06\text{pH} + 0.015\log P_{O_2} \quad [6.76]$$

These electrodes are used, in particular, as electrodes indicative of the partial oxygen pressure in a medium; they are often call *oxygen probes*.

6.4. Equilibrium of two ionic conductors in contact

We shall now examine what happens when we bring two different ionic conductors into contact with one another, thus forming an electrochemical system. Obviously, between those phases, a potential difference is established. Earlier on (in section 6.2.3), we referred to this as the *junction potential difference*.

6.4.1. *Junction potential with a semi-permeable membrane*

We shall now examine the case of two potassium chloride solutions at different concentrations C_1 and C_2. The two phases are separated by a semi-permeable membrane which allows only the potassium ions to pass through. In such a case, a genuine potential difference of thermodynamic origin is established. We can formulate that potential difference by writing the equality of the electrochemical potentials of a given ion (that which crosses the membrane – in our example, potassium):

$$\mu^0_{K^+} + RT\ln C_1 + \mathfrak{F}\Phi_1 = \mu^0_{K^+} + RT\ln C_2 + \mathfrak{F}\Phi_2 \quad [6.77]$$

We then find:

$$\Phi_1 - \Phi_2 = \frac{RT}{\mathcal{F}} \ln \frac{C_2}{C_1} \qquad [6.78]$$

That potential difference is called the *membrane potential difference* or *Donnan potential difference.*

6.4.2. Junction potential of two electrolytes with a permeable membrane

We now place our two potassium chloride solution in contact with one another, but separated, this time, by a membrane permeable to both species of ions: chloride and potassium. By virtue of the principle of thermodynamic equilibrium, we must simultaneously have equality [6.78] and the equivalent equality for the chloride ions:

$$\mu^0_{Cl^-} + RT \ln C_1 + \mathcal{F}\Phi_1 = \mu^0_{Cl^-} + RT \ln C_2 + \mathcal{F}\Phi_2 \qquad [6.79]$$

This is possible only if $C_1 = C_2$ and $\Phi_1 = \Phi_2$. Thus, we observe diffusion until identical phases are obtained. However, during the course of that diffusion, a contact potential difference is established, which is not thermodynamic in nature, but instead due to the diffusion of ions from one electrolyte 1 to the other electrolyte 2. It is a kinetically-based potential difference which is calculated by bringing the mobility of the ions into play.

We accept Thomson's hypothesis whereby the relations of conventional thermodynamics are still applicable to reversible processes taking place in a system which is not at equilibrium but is in a steady state.

Consider an elementary layer of thickness dx in the interphase membrane. Using an external generator, we pass a charge of 1 faraday across that layer at steady-state equilibrium at constant temperature and pressure. The current transported in an electrolyte by each species of ions is proportional to its transport number. The passage of 1 faraday across the layer corresponds to the passage, in the same direction, of:

$$n_i = \frac{t_i}{z_i} \qquad [6.80]$$

ions-grams with the charge z_i and transport number t_i. The variation of the electrochemical Gibbs energy will therefore be:

$$d\tilde{G} = \sum_i n_i \, \mathrm{d}\mu_i + \mathcal{F}\,\mathrm{d}\Phi \qquad [6.81]$$

As the system is steady at constant temperature and pressure, according to Thomson's hypothesis, the variation of the corresponding electrochemical Gibbs energy should be zero, so:

$$\sum_i n_i \, \mathrm{d}\mu_i + \mathcal{F}\,\mathrm{d}\Phi = 0 \qquad [6.82]$$

From this, we deduce:

$$\mathrm{d}\Phi = -\frac{1}{\mathcal{F}}\sum_i \frac{t_i}{z_i}\mathrm{d}\mu_i = -\frac{\mathrm{R}T}{\mathcal{F}}\sum_i \frac{t_i}{z_i}\mathrm{d}(\ln a_i) \qquad [6.83]$$

To find the junction potential, we integrate for the whole of the interphase, and obtain:

$$e_{\text{junct}} = \Phi_2 - \Phi_1 = -\frac{\mathrm{R}T}{\mathcal{F}}\int_1^2 \frac{t_i}{z_i}\mathrm{d}(\ln a_i) \qquad [6.84]$$

If the two phases 1 and 2 contain a single electrolyte z,z (same electrovalence) at two different concentrations C_1 and C_2, relation [6.84] becomes:

$$e_{\text{junct}} = -\frac{\mathrm{R}T}{\mathcal{F}}\int_1^2 \frac{t_+}{z}\mathrm{d}(\ln a_1) + \frac{\mathrm{R}T}{\mathcal{F}}\int_1^2 \frac{t_-}{z}\mathrm{d}(\ln a_2) \qquad [6.85]$$

Suppose that, within the small window of variation of the activities, the transport numbers of the ions do not vary. In this case, the above relation becomes:

$$e_{\text{junct}} = -\frac{\mathrm{R}T}{z\mathcal{F}}(t_+ - t_-)\ln\frac{a_2}{a_1} \qquad [6.86]$$

If the anionic transport numbers and cationic ones are equal (near to 0.5), the junction potential is near to zero; it reaches a value of a few millivolts if the transport numbers are not hugely different, as is the case for the solution 0.1 M HCl, 0.1 M KCl ($e_{\text{junct}} = 28\text{ mV}$) or for the solution 0.2M NaCl, 0.2 M NaOH ($e_{\text{junct}} = 19\text{ mV}$), owing to the high mobilities of the H^+ and OH^- ions.

To eliminate the junction potential, we can add a spectator electrolyte which has ions of around the same dimension, and thus the same mobility and same transport number. That electrolyte is added into both compartments at the same concentration. If that concentration is much higher than that of all the other ions, this electrolyte will assume responsibility for almost all of the transport of electrical current, and there will be no junction potential.

In fact, the added electrolyte can greatly influence the activity of the other ions in media 1 and 2. For this reason, it is preferable to link media 1 and 2 by a siphon filled with a concentrated potassium chloride solution, and in this case, the sum of the two junction potentials created is practically zero.

NOTE.– If, in relation [6.86], for very dilute solutions, we treat the activities and concentrations as being identical, we obtain the relation:

$$\Phi_2 - \Phi_1 = -\frac{\text{R}T}{z\mathcal{F}}(t_+ - t_-)\ln\frac{C_2}{C_1} \qquad [6.87]$$

This simplified relation is known as the Planck–Henderson relation.

6.5. Applications of Nernst's relation to the study of various reactions

We shall now demonstrate the usefulness of Nernst's relation in a number of cases such as: the prediction of redox reactions, the explanation of dismutation, redox catalysis or induced oxidation.

6.5.1. *Prediction of redox reactions*

Imagine we want to predict the direction of progression of the reaction [6R.16] between an oxidant and a reductant, written in the form:

$$\nu_2\text{Ox}_1 + \nu_1\text{Red}_2 = \nu_2\text{Red}_1 + \nu_1\text{Ox}_2 \qquad [6\text{R}.16]$$

This reaction is accompanied by a Gibbs energy $\Delta_{16}G$.

We can break down the reaction under study into two redox couples whose Gibbs energies are, respectively, $\Delta_{17}G$ and $\Delta_{18}G$.

$$Ox_1 + \nu_1 e = Red_1 \qquad [6R.17]$$

$$Ox_2 + \nu_2 e = Red_2 \qquad [6R.18]$$

The Gibbs energy of the reaction studied here is written as a function of the Gibbs energies of the reactions [6R.17] and [6R.18] thus:

$$\Delta_{16}G = \nu_2 \Delta_{17}G - \nu_1 \Delta_{18}G \qquad [6.88]$$

If, for each of these reactions, we use relation [6.44], then as a function of their electrode potentials, we obtain:

$$\Delta_{16}G = -\nu_1 \nu_2 \mathfrak{F}\left(e_{(17)} - e_{(18)}\right) \qquad [6.89]$$

We know that it is possible for reaction [6R.16] to take place from left to right only if the corresponding Gibbs energy $\Delta_{16}G$ is negative.

However, we have $\Delta_{16}G \leq 0$ if $e_{(1)7} \geq e_{(18)}$. Otherwise, it would be the reverse reaction which is possible.

Thus, by comparing the two redox potentials, we are able to predict the possible direction of the reaction. The couple with the greater potential reacts over the couple with the lesser potential.

6.5.2. Relations between the redox voltages of different systems of the same element

A very great many elements have several degrees of oxidation, and may therefore participate in multiple redox systems. Therefore, there is a relation between the normal voltages of those couples, which is known as *Luther's formula*. The diagram introduced by Frost is very useful in instantly recognizing the stability of the different degrees of oxidation of the same element and any dismutation thereof.

6.5.2.1. *Luther's formula*

Consider an element A which has three degrees of oxidation d_1, d_2 and d_3 such that $d_1 < d_2 < d_3$. The corresponding redox systems are represented by reactions [6R.19], [6R.20] and [6R.21].

$$A^{d_2} + (d_2 - d_1)\ e = A^{d_1} \qquad (e_{1,2}) \qquad [6R.19]$$

$$A^{d_3} + (d_3 - d_2)\ e = A^{d_2} \qquad (e_{2,3}) \qquad [6R.20]$$

$$A^{d_3} + (d_3 - d_1)\ e = A^{d_1} \qquad (e_{1,3}) \qquad [6R.21]$$

Depending on the nature of the species A^{d_1}, A^{d_2} and A^{d_3} involved, H^+ ions and water molecules may play a part in these reactions.

In any case, the last reaction is the sum of the first two. Hence, the Gibbs energies will obey the relation:

$$\Delta_{1.3}G = \Delta_{1.2}G + \Delta_{2.3}G \qquad [6.90]$$

By applying relation [6.44] for each of these Gibbs energies, after simplification we obtain:

$$(d_3 - d_1)\, e_{1.3} = (d_2 - d_1)\, e_{1.2} + (d_3 - d_2)\, e_{2.3} \qquad [6.91]$$

Equation [6.91] constitutes Luther's formula for three given degrees of oxidation of the same element. This relation does not depend on the concentrations and is therefore also valid for the normal voltages.

Because we have the equality $(d_3 - d_1) = (d_3 - d_2) + (d_2 - d_1)$, Luther's formula dictates that the voltage $e_{1,3}$ always lies between the voltages $e_{1,2}$ and $e_{2,3}$.

As nothing imposes an order between $e_{1.2}$ and $e_{2.3}$, it follows that we encounter two cases:

– if:

$$e_{1.2} < e_{1.3} < e_{2.3} \qquad [6.92]$$

the degree d_2 is stable, and we observe the anti-dismutation reaction:

$$(d_3 - d_2)\mathrm{A}^{d_1} + (d_2 - d_1)\mathrm{A}^{d_3} = (d_3 - d_1)\mathrm{A}^{d_2} \qquad [6R.22]$$

– if:

$$e_{2.3} < e_{1.3} < e_{1.2} \qquad [6.93]$$

the degree d_2 is unstable, and we observe the opposite reaction to the former – i.e. the dismutation reaction.

$$(d_3 - d_1)\mathrm{A}^{d_2} = (d_3 - d_2)\mathrm{A}^{d_1} + (d_2 - d_1)\mathrm{A}^{d_3} \qquad [6R.23]$$

By comparing reactions [6R.22] and [6R.23] respectively with reactions [6R.9] and [6R. 4], we see that the first case is that of iron, whilst the second is that of copper.

6.5.2.2. *Frost diagram*

The Frost diagram is plotted for an element which exhibits several different degrees of oxidation. In the Frost diagram, we place the products $de_{0,d}$ of the degree of oxidation of the compound by the potentials of the couples formed by the element (degree 0) and the other degrees of oxidation in a system of axes, with $e_{0,d}$ on the ordinate axis and the degrees of oxidation d on the abscissa. We use either the normal voltages or the real voltages with given concentrations. Each degree of oxidation (other than 0) is thus represented by a point in the diagram (Figure 6.6). For the degree 0, as the potential $e_{0.0}$ is not defined, it is agreed to represent it as the origin of the axes.

Consider two points in that diagram. The slope of the straight line connecting those points is the voltage e_{d_1,d_2}. Indeed, by calculating the slope, we obtain:

$$p = \frac{d_2 e_{0,d_2} - d_1 e_{0,d_1}}{d_2 - d_1} \qquad [6.94a]$$

and by applying Luther's relation [6.91], we do indeed find the potential e_{d_1,d_2}:

$$p = e_{d_1,d_2} \qquad [6.94b]$$

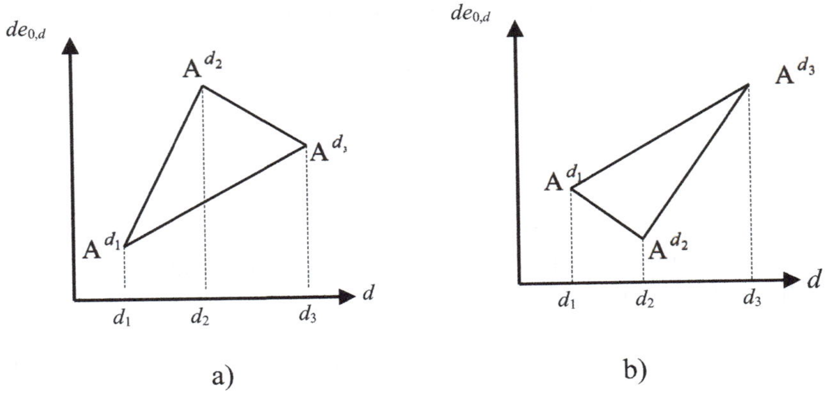

Figure 6.6. *Frost diagram for three degrees of oxidation of an element*

In order to plot the Frost diagram, obviously we need to know all the potentials $e_{0,d}$. Some of them are found in the tables; others are calculated on the basis of other potentials, or by application of Luther's relation [6.91].

Look at Figure 6.6, representing three degrees of oxidation of an element: d_1, d_2 and d_3. The arrangement of these three points corresponds either to Figure 6.6(a) or Figure 6.6(b). In the first case, the degree of oxidation d_2 is not stable and is dismuted. In the second case, however, we observe anti-dismutation between d_1 and d_3.

It is possible to work with forms of the same degree of oxidation of an element, linked by an acid–base reaction or a hydration reaction, such as CuO and $Cu(OH)_2$, in which case the two compounds lie on the same vertical line. Thus, two forms of the same element, represented by two points on the same vertical in the Frost diagram, in fact belong to the same degree of oxidation and are connected to one another by an acid–base reaction or a dehydration reaction.

Table 6.1, for example, shows the degrees of oxidation of manganese and the different corresponding chemical forms. Figure 6.7 is the representation of the Frost diagram for manganese, calculated at a pH of 0 with the activities of the other species in solution being equal to 1.

Degree of oxidation	Chemical form	Point in Figure 6.6
0	Mn metal	A
2	Mn^{2+} (manganous ion)	B
3	Mn^{3+} (manganic ion)	C
4	MnO_2 (dioxide)	D
6	MnO_4^{2-} (manganate ion)	E
7	MnO_4^- (permanganate ion)	F

Table 6.1. *The different degrees of oxidation of manganese*

The polygonal line ABDF is such that all the points in the diagram are either on or above that line. It is known as the polygon of stability. The peaks of that line correspond to stable compounds; points below axis are unstable, and hence liable to break down either by dismutation or by an acid–base reaction or dehydration reaction.

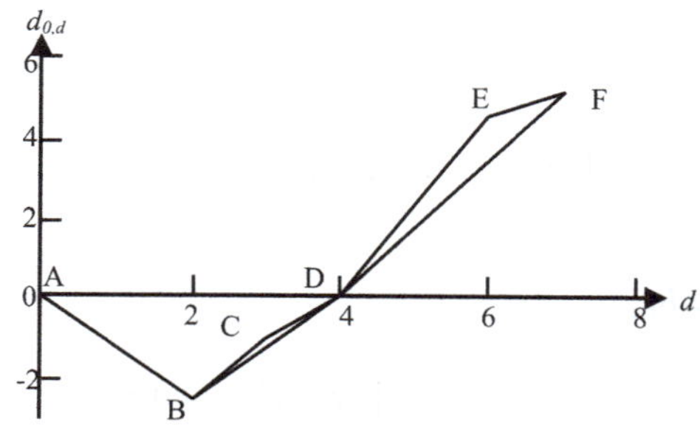

Figure 6.7. *Frost diagram of manganese at pH = 0*

6.5.3. *Predicting the dismutation and anti-dismutation reactions*

A dismutation reaction occurs when a compound A is likely to belong to two redox systems:

– one system in which A is the oxidant;

– one system in which A is the reductant.

Such is the case, for instance, with the copper ion Cu^{+}, which plays that double role in to the two redox systems. In one of the systems, the copper ion is an oxidant and works thus:

$$Cu^{+} + e = Cu \qquad (e_{12}^{0} = 0.52V) \qquad [6R.24]$$

In the other system, the copper ion plays the role of reductant, and reacts thus:

$$Cu^{++} + e = Cu^{+} \qquad (e_{13}^{0} = 0.16V) \qquad [6R.25]$$

The system [6R.24], therefore, is able to reduce the system [6R.25], so we would observe the reaction:

$$2Cu^{+} = Cu^{2+} + Cu \qquad [6R.26]$$

This is a dismutation reaction, which will continue until an extremely low concentration of copper ions is obtained.

The reverse reaction, which we see, for instance, with ferric ions, and is written:

$$Fe + 2Fe^{3+} = 3Fe^{2+} \qquad [6R.27]$$

is an example of an anti-dismutation reaction.

6.5.4. *Redox catalysis*

Look again at the reaction [6R.16], which occurs spontaneously from left to right. In order to speed up this reaction, we use a catalyst, present in two different forms: one oxidized (Oxcat) and the other reduced (Redcat). Reaction [6R.16] would therefore by replaced by two redox reactions: [6R.28] and [6R.29]:

$$\nu_{cat}\mathrm{Ox}_1 + \nu_1\mathrm{Red}_{cat} = \nu_{cat}\mathrm{Red}_1 + \nu_1\mathrm{Ox}_{cat} \qquad [6R.28]$$

$$\nu_2\mathrm{Ox}_{cat} + \nu_{cat}\mathrm{Red}_2 = \nu_2\mathrm{Red}_{cat} + \nu_{cat}\mathrm{Ox}_2 \qquad [6R.29]$$

These two reactions involve the redox couple of the catalyst, which is written:

$$\mathrm{Ox}_{cat} + \nu_{cat}\mathrm{e} = \mathrm{Red}_{cat} \qquad [6R.30]$$

In order for the system to work – i.e. in order for both reactions [6R.28] and [6R.29] to occur spontaneously – it is easy to show that the three potentials must satisfy the twofold inequality:

$$e_{28} > e_{30} > e_{29} \qquad [6.95]$$

The potential of the catalyst couple must have a value between those of the potentials of the two couples in the desired reaction.

6.6. Redox potential in a non-aqueous solvent

All our discussions in Chapter 4 about redox reactions in aqueous media can immediately be transposed to apply to non-aqueous solvents, although we need to take account of the ion pairs for solvents with a low dielectric constant, as we did in Chapter 3.

In a given solvent, it is obviously necessary to establish a scale of the potentials of the redox couples in that solvent.

6.6.1. *Scale of redox potential in a non-aqueous medium*

In order to establish a scale of redox potential in a solvent, we need to have a reference electrode in that medium.

Whatever the medium, a reference electrode needs to be chosen on the basis of a number of electrochemical and technological criteria.

On the electrochemical level, the reference electrode must have the following qualities:

– an electromotive force (emf) which is definite and stable over time;

– an emf independent of any chemical modifications of the medium;

– good reversibility and low polarizability;

– good reproducibility;

– compatibility with the temperature range.

On the technological level, the criteria needing to be taken into account are:

– easy implementation and use;

– risk of contamination from the bath reduced to the minimum;

– stable junction potentials (usually negligible in the case of porous junctions).

6.6.1.1. *Reference electrodes in a non-aqueous molecular solvent*

For reference electrodes in a non-aqueous medium, we distinguish between two cases: molecular solvents and ionic solvents.

When possible, the hydrogen electrode in the solvent in question (pH(S) = 0 and $P_{H2} = 1$ bar, like in the aqueous phase, is preferred.

If that electrochemical equilibrium is not feasible, any other system exhibiting the required qualities can be used. For example, often, wherever possible, we use the following electrode:

Pt/(ferrocene + ferricinium$^+$ cation) in S

Ferrocene is a molecular complex between an Fe^{2+} ion and two cyclopentadienyl anions. By loss of an electron, this complex transforms into a ferricinium cation with the same structure with the positive charge distributed symmetrically. Thus, we have the equilibrium:

ferricinium$^+$ + e = ferrocene [6R.31]

6.6.1.2. *Reference electrodes in an ionic solvent (molten media)*

In a molten-salt medium, gas electrodes, such as the hydrogen electrode, are not easy to implement, given the difficulty of managing the gas at a high temperature.

Generally speaking, the technological problems linked to high temperature greatly complicate the realization of a reference electrode in a molten medium. It is necessary to use suitable materials, but they are not always easy to work with.

In view of the diversity of potential situations, there is no universal reference electrode in molten media.

Figures 6.7(a) and 6.7(b) show two examples of electrodes that can be used at high temperature.

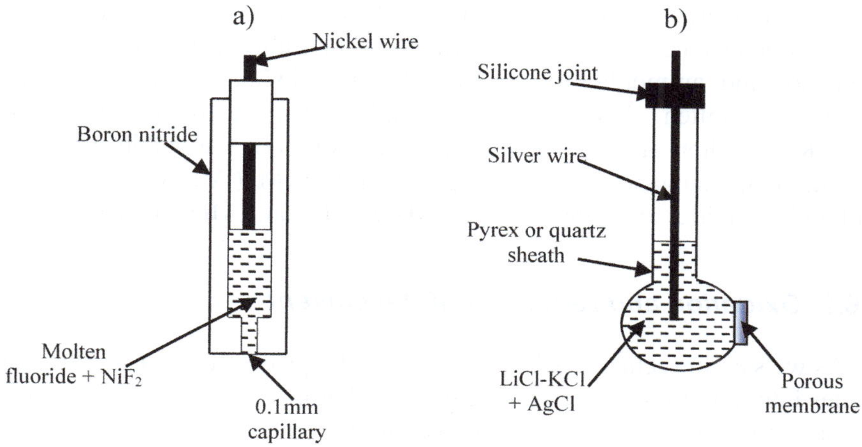

Figure 6.8. *Reference electrodes in a molten medium: a) reference electrode in a fluoride medium; b) reference electrode in a chloride medium (from [GUE 09])*

Figure 6.8(a) represents a liquid-junction NiF_2/Ni electrode usable in a fluoride medium. Note that the casing is made of boron nitride – a material very similar to carbon.

Figure 6.8(b) represents a molten AgCl/Cl electrode for the chloride medium with a porous membrane of quartz, yttria-stabilized zirconia or aluminum oxide.

Once a reference electrode has been chosen, then in relation to that reference electrode, we determine the normal potentials of the redox couples existing in the solvent being used.

We then note highly different redox properties from one solvent to another.

The normal potential of a given redox couple often differs from one solvent to another. For instance, that of the couple Cu^{2+}/Cu, which is 0.34 V in water, becomes 0.28 V in formamide and 0.34 V in the equimolecular mixture of KCl and molten NaCl at 700°C.

Certain degrees of oxidation are stable in certain solvents but not in others. For instance, the Cu^{+} ion, which is not stable in water, is stable in methanol and ammonia. It is also stable in ethylammonium chloride at 127°C, in molten alkali hydroxides and in the equimolecular mixture NaCl/KCl molten at 700°C. In an aqueous solution, the stable degrees of oxidation of manganese are the degrees Mn(II), Mn(IV) and Mn(VII). In molten hydroxides, only the states Mn(II), Mn(III) and Mn(V) are stable.

6.6.2. *Oxidation and reduction of the solvent*

As we saw in section 6.3.1.3, in the case of aqueous solutions, the solvent comes into play to limit the voltage domain that is accessible. We often have the possibility of both reduction and oxidation of the solvent.

Amphiprotic solvents, as does water, yield a reduction system which forms hydrogen gas and the anion of the conjugate strong base of the solvent:

$$2SH + 2e = H_2 + 2S^- \qquad [6R.32]$$

At the other end of the voltage range, we have the reaction of oxidation of the solvent – e.g.:

$$2SH = S_2H^- + 2e + H^+ \qquad [6R.33]$$

The breadth of the accessible voltage range depends on the solvent. Thus, in acetonitrile, it is possible to oxidize the perchlorate ion ClO_4^- into the radical $ClO_4^{\bullet}$.

When oxidized, carboxylic acids give the Kolbe reaction – i.e. decarboxylation and formation of hydrocarbons and carbon dioxide as follows:

$$2RCOO^{-} = 2e + CO_2 + RH \qquad [6R.34]$$

In ionic solvents, the limitations are due to the oxidation or reduction, either of the anion or of the cation.

For example, in molten alkali halides, the reduction of the cation to the metal limits the domain to reductant potentials, thus:

$$Na^{+} + e = Na \qquad [6R.35]$$

On the other hand, in molten alkali nitrates, it is the reduction of the nitrate anion to form a nitrite anion which limits the domain, because that reduction is easier than that of the alkali cations.

$$NO_3^{-} + 2e + 2H^{+} = NO_2^{-} + H_2O \qquad [6R.36]$$

6.6.3. *Influence of solvent on redox systems in a non-aqueous solvent*

As we saw in section 6.2.5 in the case of water, species deriving from the solvent may play a part in the redox reaction of certain couples. The reaction involves these species as well as the electrons. In the case of protic solvents, one of the essential variables, as we saw with water, is the pH. In the case of non-protic acidity, then as we did in Figure 6.1, we plot potential/acidity curves, with the acidity being characterized by pCl^{-} if we are dealing with chloroacidity, pO^{2-} for oxy-acidity or pH_2O in molten alkali hydroxides (see section 3.12).

7

Thermodynamics of Electrochemical Cells

Clearly, in practice, it is not possible to use just one electrode, because then we are unable to find the absolute potentials, and only the potential differences can be measured and used. Therefore, it is always essential to use two electrodes in combination to constitute what is known as a galvanic cell, which forms a new electrochemical system.

7.1. Electrochemical chains – batteries and electrolyzer cells

In fact, the two electrodes in a galvanic cell may be separated by any number of junction electrolytes, with the whole forming an electrochemical chain.

The phases at the ends of the chain, known as the *poles*, are always composed of the same electron conductor, which is generally a metal but also sometimes a semi-conductor, whilst the intermediary phases are ion conductors – usually liquid but occasionally solid.

Thus, we represent a galvanic chain by writing the succession of phases separated by a vertical or inclined bar, a double bar representing a diaphragm or membrane between two liquid phases. In practice, we do not write the second metal end phase, which is identical to that which is at the other end. Thus, the first voltaic cell (named for its inventor, Alessandro Volta) is a chain with only one electrolyte, represented by:

Cu/aqueuos solution H_2SO_4 | Zn

Of course, a copper wire is, it goes without saying, connected to the zinc.

The Daniell cell, which is more complex, contains two electrolytic solutions separated by a membrane and is represented by the chain:

$$Cu \mid Cu^{++}(\text{aqueuous solution } CuSO_4) \parallel Zn^{++}(\text{aqueous solution } ZnSO_4) \mid Zn$$

Such chains, where the endpoints are different, are called *polar chains*, by contrast to the following chain:

$$Cu \mid Cu^{++}(\text{aqueous solution } CuSO_4) \mid Cu$$

which is an apolar chain.

If a polar chain is placed in an electrical circuit, meaning that the end two electron conductors are connected by a metal conductor, two phenomena may occur:

– either a current passes spontaneously in the circuit, in which case the chain constitutes a generator. We say then that the electrolytic chain is a *battery*;

– or else it is necessary to place an electricity generator in the external circuit for the current to pass into the chain. The chain then functions as a receiver, and we say that we are dealing with an *electrolyzer* or an *electrolysis cell.*

NOTE.– When we are interested only in the thermodynamics of the chain at equilibrium, there is no current and there is no longer any distinction between a battery and an electrolyzer. It is for this reason that, in this case, the term *cell* replaces the term *galvanic cell at equilibrium*, whatever its true nature. It is also common to speak of an *electrochemical cell.*

7.2. Electrical voltage of an electrochemical cell

Consider a very general chain whose phases are number 1 to *n*:

$$1 \mid 2 \mid 3 \mid 4 \ldots\ldots\ldots n-1 \mid n \mid 1'$$

The chain ends with the phase 1', which is of the same nature as the metal phase 1.

The term *reversible electrical voltage* of a cell is used to denote the difference:

$$E_{\mathrm{abs}} = \Phi_1 - \Phi_{1'} \quad [7.1]$$

between the internal potentials of the extreme phases. Such a voltage may be positive or negative, and its sign changes if we invert the order in which the chain is written. The positive pole is that which has the highest voltage, and by convention, we always write that pole on the left in the representation of the chain. Hence, the value E_{abs} will always be positive.

NOTE.– This potential difference is an electromotive force (emf) in the case of a battery and a counter-electromotive force (cemf) in the case of an electrolyzer.

The voltage of the chain is, in fact, the sum of the differences between the potentials located at the different interphases of the chain. Thus, we can write:

$$E_{\mathrm{abs}} = (\Phi_1 - \Phi_2) + \sum_{i=2}^{n-2}(\Phi_i - \Phi_{i+1}) + (\Phi_{n-1} - \Phi_n) + (\Phi_n - \Phi_{1'}) \quad [7.2]$$

In this expression, the differences $\Phi_1 - \Phi_2$ and $\Phi_{n-1} - \Phi_n$ are the *absolute voltages* of the electrodes 1/2 and $n/n{-}1$, the differences $\Phi_i - \Phi_{i+1}$ are *junction potential differences* and the difference $\Phi_n - \Phi_{1'}$ is the *contact potential difference* between two metals.

Using expressions [6.34] and [6.35] to express those voltages and taking account of the fact that phases 1 and 1' are identical, we obtain the expression:

$$E_{\mathrm{abs}} = e_{\mathrm{rel}}^{+} - e_{\mathrm{rel}}^{-} + \sum_j \varepsilon_j \quad [7.3]$$

Here, e_{abs}^{+} and e_{abs}^{+} are the relative potentials of the positive pole and the negative pole, and $\sum_j \varepsilon_j$ is the sum of the different junction potentials and contact potentials. This latter term has a value of 0 if there is only one electrolyte. In this simple case, the above relation is simplified to give:

$$E_{\text{abs}} = e^{+}_{\text{rel}} - e^{-}_{\text{rel}} \qquad [7.4]$$

NOTE.– The difference between the absolute potentials is identical to the difference between the relative potentials if we use the same reference electrode, such as the hydrogen electrode, to evaluate those relative potentials. Thus, we can write relations [7.3] and [7.4] in the forms:

$$E_{\text{abs}} = e^{+} - e^{-} + \sum_{j} \varepsilon_{j} \qquad [7.5]$$

$$E_{\text{abs}} = e^{+} - e^{-} \qquad [7.6]$$

Thus, these expressions involve the electrode potentials, found in the tables for example.

7.3. Cell reaction

Consider a cell, which is composed of two electrodes, each of which is characterized by an electrode reaction. The "cell reaction" is the linear combination of the two electrode reactions which eliminates the electrons.

For example, if we look again at the Daniell cell, represented by the chain:

$$\text{Cu} \,|\, \text{Cu}^{++} \,\|\, \text{Zn}^{++} \,|\, \text{Zn}$$

at the positive pole of the copper electrode, the addition of electrons is expressed by the reaction:

$$\text{Cu}^{++} + 2\text{e} = \text{Cu} \qquad [7\text{R}.1]$$

At the negative pole of the zinc electrode, the electrode reaction, which yields electrons, is written:

$$\text{Zn} = 2\text{e} + \text{Zn}^{++} \qquad [7\text{R}.2]$$

The linear combination to eliminate the electrons, here, is a simple sum, and the cell reaction would therefore be:

$$Cu^{++} + Zn = Cu + Zn^{++} \qquad [7R.3]$$

Of course, the thermodynamic functions such as the enthalpy, entropy or Gibbs energy of a cell reaction can be deduced from the corresponding functions of the electrode reactions by the same linear combination.

One particular family of cells uses the reaction of combustion of a combustible gas such as hydrogen, methane, etc. In these cells, known as *fuel cells* (FCs), the positive electrode is an oxygen electrode which works as follows:

$$\frac{1}{2}O_2 + 2e = O^- \qquad [7R.4]$$

The negative electrode is also a gas electrode, in which the gas reacts with the oxygen ions to yield electrons – for instance, with carbon monoxide, the electrode reaction is:

$$CO + O^{2-} = CO_2 + 2e \qquad [7R.5]$$

The cell reaction then appears to be the reaction of combustion of the gas. The oxygen ions are transported from one electrode to the other through a solid electrolyte. It is common to use zirconia (ZrO_2), which is a vacancy conductor of oxygen ions, in which we increase the number of anionic vacancies by doping it with a trivalent cation, such as yttrium. The cell functions at a high temperature (800°C).

7.4. Influence of temperature on the cell voltage; Gibbs–Helmholtz formula

Based on the general relation [6.45] applied to the cell voltage, and relation [6.59], we can immediately deduce the two relations:

$$E_{abs} = \frac{\Delta_{Rc}H}{\nu_e \mathfrak{F}} - \frac{T}{\nu_e \mathfrak{F}}\left(\frac{\partial E_{abs}}{\partial T}\right)_{P,n_i} \qquad [7.7]$$

$$\left(\frac{\partial E_{\text{abs}}}{\partial T}\right)_{P,n_i} = -\frac{\Delta_{Rc} S}{\nu_e \mathcal{F}} \qquad [7.8]$$

In these relations, ν_e is the number of electrons exchanged between the two electrodes – i.e. the lowest common multiple of the numbers of electrons involved in each of the reactions of the electrodes. $\Delta_{Rc} H$ and $\Delta_{Rc} S$ are, respectively, the enthalpy and entropy associated with the cell reaction *Rc*. The differential $\left(\frac{\partial E_{\text{abs}}}{\partial T}\right)_{P,n_i}$ is called the cell's temperature coefficient.

The first of these relations is a relation similar to Helmholtz's law for the Gibbs energy.

7.5. Influence of activity on the cell voltage

Using Nernst's relation [6.42] for each electrode reaction, and combining the two relations obtained by way of the same linear combination as that used to obtain the cell reaction, we immediately obtain the relation between the cell voltage and the activities of the reagents and the products of the electrode reaction. We can illustrate this on the following electrochemical chain:

$$\text{Pt} \mid \text{Cl}_2 \text{ (gas)} \mid \text{Zn}^{++} \text{(aqueous solution of zinc chloride)} \mid \text{Zn}$$

The electrode reactions are, respectively, for the positive pole and negative pole:

$$\text{Cl}_2 + 2\text{e} = 2\text{Cl}^- \qquad [7R.6]$$

$$\text{Zn} = 2\text{e} + \text{Zn}^{++} \qquad [7R.7]$$

The cell reaction becomes:

$$\text{Cl}_2 + \text{Zn} = 2\text{Cl}^- + \text{Zn}^{++} \qquad [7R.8]$$

The cell voltage would be:

$$E_{\text{abs}} = e^{(+)} - e^{(-)} = e^0_{Cl} - e^0_{Zn} + \frac{RT}{2\mathfrak{F}} \ln P_{Cl_2} - \frac{RT}{2\mathfrak{F}} \ln \left|Zn^{++}\right| \left|Cl^-\right|^2 \qquad [7.9]$$

This relation involves only the product $\left|Zn^{++}\right| \left|Cl^-\right|^2$, rather than the individual activity of the ions. By finding the product of the mean concentration (defined by relation [1.3]) by the mean activity coefficient (defined by relation [1.1]), we obtain the mean activity, which, in our case, is:

$$\left|ZnCl_2\right|_\pm = \left(\left|Zn^{++}\right| \left|Cl^-\right|^2\right)^{1/2} \qquad [7.10]$$

The voltage of the chain becomes:

$$E_{\text{abs}} = e^0_{\text{Cl}} - e^0_{\text{Zn}} + \frac{RT}{2\mathfrak{F}} \ln P_{\text{Cl}_2} - \frac{3RT}{2\mathfrak{F}} \ln \left|ZnCl_2\right|_\pm \qquad [7.11]$$

As we can see, by experimentally measuring the cell voltage, we are able to find the mean activities.

The difference $E^0 = e^0_- - e^0_+$ is called the *standard potential* of the cell, and relation [7.11] is then written:

$$E_{\text{abs}} = E^0 + \frac{RT}{2\mathfrak{F}} \ln P_{\text{Cl}_2} - \frac{3RT}{2\mathfrak{F}} \ln \left|ZnCl_2\right|_\pm \qquad [7.12]$$

We shall see later on (in section 7.7.1) how to directly measure that potential if the electrode potentials e^0_- and e^0_+ are unknown.

7.6. Dissymmetry of cells, chemical cells and concentration cells

The calculation of the cell voltage in section 7.2 and relation [7.6] clearly demonstrate that the condition necessary to obtain an electrochemical cell is that there be dissymmetry in the chain of electrolytes, so that the voltage in non-null. This dissymmetry may be produced in a variety of ways.

As we saw earlier, it is possible to create dissymmetry by using two chemically-different electrodes. In this case, we obtain *chemical cells*.

A second way of creating dissymmetry is to place the same electrolyte at the extremities of the cell, but at different concentrations – e.g. for the chain:

$$\mathrm{Cu} \mid \mathrm{Cu}^{2+}(\text{concentration } C_1) \parallel \mathrm{Cu}^{2+}(\text{concentration } C_2) \mid \mathrm{Cu}^{2+} \mid \mathrm{Cu}$$

With concentrations $C_1 > C_2$, we obtain a *concentration cell*. The voltage of such a cell would be given, if we overlook the junction voltage between the two electrolytes and treat the concentration and activity as being the same, by Nernst's relation:

$$\mathrm{E}_{\mathrm{abs}} = \frac{\mathrm{R}T}{2\mathcal{F}} \ln \frac{C_1}{C_2} \qquad [7.13]$$

In this case, the cell reaction would be the simple passage of the copper solution from the concentration C_1 to the concentration C_2, which can be written in the form:

$$\mathrm{Cu}^{2+}(C_1) = \mathrm{Cu}^{2+}(C_2) \qquad [7R.9]$$

A particular case of concentration cells is a cell where the difference in concentration is due to a field of external force, such as the field of gravity. We then obtain what is known as a *gravity cell*.

7.7. Applications to the thermodynamics of electrochemical cells

Measurements of the voltage of electrochemical cells at equilibrium are at the root of the methods used to determine a great many thermodynamic variables. What follows is a breakdown of the main methods, classified on the basis of the thermodynamic value sought.

7.7.1. *Determining the standard potentials of cells*

The standard cell potential can be measured directly on the basis of the measurements of the cell's emf at different electrolyte concentrations.

We shall demonstrate the method using the example of measuring the standard potential of a zinc chloride cell. For this purpose, we consider the following junctionless cell:

Zn | ZnCl2 (C) | AgCl,Ag

The cell voltage would be:

$$E_{abs} = E^0 - \frac{RT}{2\mathfrak{F}} \ln \left|Zn^{++}\right| \left|Cl^-\right|^2 \quad [7.14]$$

This relation involves only the product of the activities $\left|Zn^{++}\right| \left|Cl^-\right|^2$, rather than the individual activity of the ions.

$$\left|Zn^{++}\right| = [Zn]\gamma_{Zn^{++}} = C\gamma_{Zn^{++}} \quad [7.15a]$$

and

$$\left|Cl^-\right| = \left[Cl^-\right]\gamma_{Cl^-} = 2C\gamma_{Cl^-} \quad [7.15b]$$

The voltage of the chain becomes:

$$E_{abs} = E^0 - \frac{RT}{2\mathfrak{F}} \ln 4C^2 - \frac{RT}{2\mathfrak{F}} \ln \gamma_{Zn^{++}} \gamma_{Cl^-}^2 \quad [7.16]$$

However, the mean activity coefficient is defined by:

$$\gamma_\pm^3 = \gamma_{Zn^{++}} \gamma_{Cl^-}^2 \quad [7.17]$$

Thus, relation [7.3] can be written in the form:

$$E_{abs} + \frac{RT}{2\mathfrak{F}} \ln 4C^2 = E^0 - \frac{3RT}{2\mathfrak{F}} \ln \gamma_\pm \quad [7.18]$$

We measure the potential E_{abs} for different concentrations C, and express the quantity $E_{abs} + \frac{RT}{2\mathfrak{F}} \ln 4C^3$ as a function of the concentration C (Figure 7.1). By extrapolation to concentrations of 0, with the mean activity coefficient having a value of 1 in these conditions, the quantity $E_{\text{abs}} + \frac{RT}{2\mathcal{F}} \ln 4C^2$ has the value of E^0, according to relation [7.5].

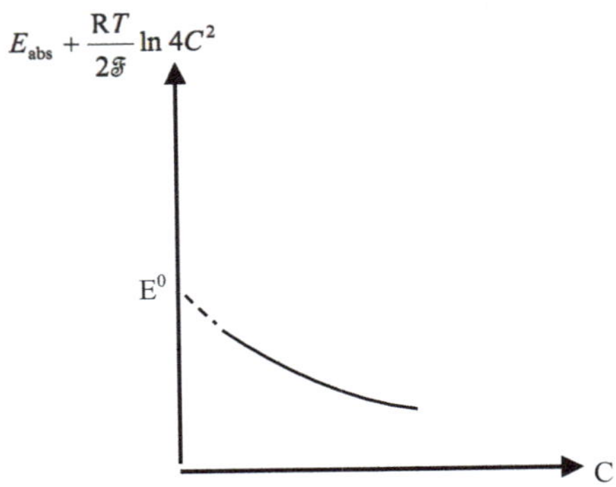

Figure 7.1. *Determination of the standard cell potential*

This method can be used to measure the standard potential of an electrode. The standard cell potential is linked to the standard potentials of the two electrodes by the relation:

$$E^0 = e_-^0 - e_+^0 \tag{7.19}$$

Thus, by measuring the standard cell potential, if the standard potential for one of the electrodes is known, then it is easy to deduce the other.

7.7.2. Determination of the dissociation constant of a weak electrolyte on the basis of the potential of a cell

We shall now describe a method for determining the dissociation constant of a weak electrolyte, by successive approximations, based on the measurement of an appropriate cell's potential. In describing this method, we shall base our discussions on the example of ethanoic acid, written as EtH. We begin by constructing the following junctionless cell:

$$\text{Pt (s)} \mid \text{H}_2 \text{ (gas, P} = 1\text{bar)} \mid \text{EtH } (C_2) \text{ EtNa } (C_3) \text{ NaCl } (C_4) \mid \text{AgCl (s)} \mid \text{Ag (s)}$$

The liquid phase is an aqueous solution of ethanoic acid (at concentration C_2), sodium ethanoate (at concentration C_3) and sodium chloride (at concentration C_4).

The electromotive force of that cell is given by the following expression, and is a function only of the activity of the chlorine- and hydrogen ions:

$$E_{abs} = E^0 - \frac{\mathrm{R}T}{\mathcal{F}}\ln\left\{\left|\mathrm{H_3O^+}\right|\;\left|\mathrm{Cl^-}\right|\right\} = E^0 - \frac{\mathrm{R}T}{\mathcal{F}}\ln\left\{\left[\mathrm{H_3O^+}\right]\;\left[\mathrm{Cl^-}\right]\gamma_{\mathrm{Cl^-}}\gamma_{\mathrm{H^+}}\right\} \qquad [7.20]$$

The proton concentration depends on the equilibrium of dissociation of the ethanoic acid, whose acidity constant is given by:

$$K = \frac{\left|\mathrm{H_3O^+}\right|\;\left|\mathrm{Et^-}\right|}{\left|\mathrm{EtH}\right|} = \frac{\left[\mathrm{H_3O^+}\right]\;\left[\mathrm{Et^-}\right]}{\left[\mathrm{EtH}\right]}\frac{\gamma_{\mathrm{H_3O^+}}\gamma_{\mathrm{Et^-}}}{\gamma_{\mathrm{EtH}}} \qquad [7.21]$$

By drawing the concentration of protons from this relation and substituting back into relation [7.20], we obtain:

$$E_{abs} = E^0 - \frac{\mathrm{R}T}{\mathcal{F}}\ln\left\{K\frac{\left[\mathrm{EtH}\right]\;\left[\mathrm{Cl^-}\right]}{\left[\mathrm{Et^-}\right]}\frac{\gamma_{\mathrm{Cl^-}}\gamma_{\mathrm{EtH}}}{\gamma_{\mathrm{Et^-}}}\right\} \qquad [7.22]$$

This expression can be written in the form:

$$\frac{\mathrm{R}T}{\mathcal{F}}\ln K' = E_{abs} - E^0 + \frac{\mathrm{R}T}{\mathcal{F}}\ln\frac{\left[\mathrm{EtH}\right]\;\left[\mathrm{Cl^-}\right]}{\left[\mathrm{Et^-}\right]} = -\frac{\mathrm{R}T}{\mathcal{F}}\ln\left(K\frac{\gamma_{\mathrm{Cl^-}}\gamma_{\mathrm{EtH}}}{\gamma_{\mathrm{Et^-}}}\right) \qquad [7.23]$$

thus defining a pseudo-constant K' which would be the value of the acidity constant of the couple EtH/$\mathrm{Et^-}$ if the solution were perfect. Thus, the two constants K and K' are linked by:

$$K' = K\frac{\gamma_{\mathrm{Cl^-}}\gamma_{\mathrm{EtH}}}{\gamma_{\mathrm{Et^-}}} \qquad [7.24]$$

In equation [7.23], the term E^0 can either be calculated or measured. We measure E_{abs}. The concentrations of the different species are:

$$\left[\text{Cl}^-\right] = C_4;\ [\text{EtH}] = C_2 - \left[\text{H}_3\text{O}^+\right];\ \left[Et^-\right] = C_3 + \left[\text{H}_3\text{O}^+\right] \qquad [7.25]$$

In general, the proton concentration is far lower than the concentrations of non-dissociated acid and ethanoate anions. Therefore, we can estimate that proton concentration using equation [7.21], taking an arbitrary value of the constant *K*.

Let *I* denote the ionic strength of the solution, which will be written:

$$I = C_3 + C_4 + \left[\text{H}_3\text{O}^+\right] \qquad [7.26]$$

When that ionic strength tends toward 0, the activity coefficients tend toward 1, and thus we have:

$$\lim_{I \to 0} \frac{\gamma_{\text{Cl}^-} \gamma_{\text{EtH}}}{\gamma_{\text{Et}^-}} = 1 \qquad [7.27]$$

Hence, if we plot K', calculated by the first equation [7.23], as a function of the ionic strength calculated by relation [7.26], and determine the limit reached as the ionic strength tends toward 0, we obtain a new value of *K*:

$$\lim_{I \to 0} K' = K \qquad [7.28]$$

A second iteration, i.e. a new calculation of the proton concentration, can be done on the basis of this new value of *K* and relation [7.21], and the procedure is repeated anew on the basis of equation [7.26]. The operation continues to be repeated until we attain a sufficient degree of precision.

NOTE.– The difficulty with this method lies in the arbitrary choice of the first approximate value of the dissociation constant. It is possible to avoid the use of an approximate value such as this. Indeed, this value only serves to set an approximate value of the proton concentration. This determination can be replaced by a measurement of the pH of the solution at each concentration in terms of ethanoic acid, and treating the activity of the protons as being equal to their concentration, i.e. positing:

$$\left[H_3O^+\right] \approx \exp(-pH) \qquad [7.29]$$

Note that, even from the first iteration, the initial value thus calculated for the constant K is generally not too far away from the true value. A second iteration is usually sufficient.

7.7.3. *Measuring the activity of a component in a strong electrolyte*

In a strong electrolyte, we may need to measure either the absolute activity of an ion or the mean activity of the electrolyte.

7.7.3.1. *Measuring the absolute activity of an ion*

The method for determining the absolute activity of an ion in a strong electrolyte (salts, strong acids and bases) is based on the complete dissociation of that electrolyte into ions, and on the measurement of the electromotive force of a cell involving that ion. We use Nernst's law, which the cell obeys if its behavior is reversible.

Thus, it is sufficient to create a cell formed of an active electrode of the ion under study or the measuring electrode, and a reference electrode. In general, the liquids in which the electrodes are immersed are connected by a siphon filled with a concentrated electrolyte, in order to minimize the junction potentials.

For example, to measure the activity of the hydrogen ion, we can use the setup combining a calomel electrode (reference electrode) and a hydrogen electrode (measuring electrode), connected by a junction siphon (Figure 7.2).

The potential is then measured at 25°C, and obeys the relation:

$$E_{\text{abs}} = 0.252 - 0.06 \ln\left|H_3O^+\right| \qquad [7.30]$$

By measuring this potential, therefore, we are able to measure the solution's proton activity – in other words, the pH.

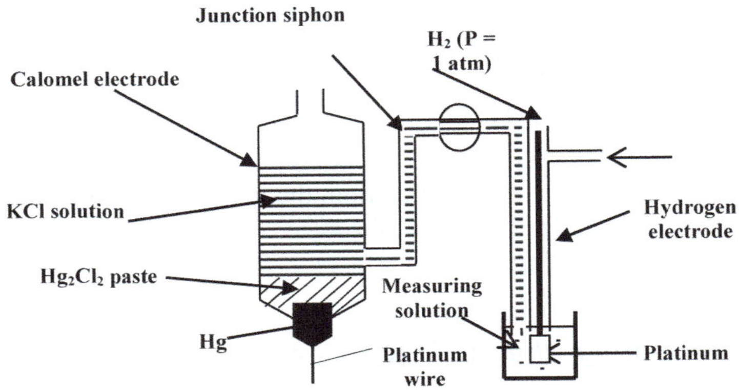

Figure 7.2. *Diagram of the calomel electrode*

A variant of this method is to create a concentration cell by taking two identical electrodes, immersed in two solutions with different activities, one of which is known.

For instance, we immerse two silver electrodes in two solutions of Ag^+ ions: one in which the activity of the silver is known (solution 2) and the other being the solution in which we wish to measure the activity of the silver ion (solution 1). The potential of the cell thus constituted would be:

$$E_{\text{abs}} = e^0_{\text{Ag}} + 0.06 \ln \left|\text{Ag}^+\right|_2 - e^0_{\text{Ag}} - 0.06 \ln \left|\text{Ag}^+\right|_1 \quad [7.31]$$

Thus:

$$E_{\text{abs}} = 0.06 \ln \frac{\left|\text{Ag}^+\right|_2}{\left|\text{Ag}^+\right|_1} \quad [7.32]$$

This relation can be used to measure the activity of the unknown solution without needing to know the value of the standard potential of silver e^0_{Ag}.

In practice, the true value of the activity is difficult to obtain, due to the error introduced by the junction siphon. Indeed, whilst our calculation supposes the junction potential to be zero, its true value is hard to find (see section 6.4.2).

7.7.3.2. Measuring the mean activity coefficient of a strong electrolyte

To obtain more accurate results, it is preferable to use cells without a junction. In this case, the method of measuring the activity coefficients of strong electrolytes simply gives us the mean activity coefficient.

If we know the standard potential of a carefully-selected cell (which can be determined experimentally (see section 7.7.1) or calculated using the electrode standard potential tables), then we can calculate the mean activity coefficient at any given concentration.

We shall illustrate this method using the example of measuring the mean activity coefficient of a zinc chloride solution of concentration C. With this purpose, we create the following junctionless cell:

Zn | $ZnCl_2$ (crystallized) | AgCl (dissolved), Ag

The cell voltage would be:

$$E_{abs} = E^0 - \frac{RT}{2\mathcal{F}} \ln\left|Zn^{++}\right| \left|Cl^-\right|^2 \qquad [7.33]$$

This relation involves only the product of the activities $\left|Zn^{++}\right| \left|Cl^-\right|^2$, rather than the individual activity of the ions.

$$\left|Zn^{++}\right| = [Zn]\gamma_{Zn^{++}} = C\gamma_{Zn^{++}} \qquad [7.34a]$$

and

$$\left|Cl^-\right| = \left[Cl^-\right]\gamma_{Cl^-} = 2C\gamma_{Cl^-} \qquad [7.34b]$$

The voltage of the chain becomes:

$$E_{abs} = E^0 - \frac{RT}{2\mathcal{F}} \ln 4C^2 - \frac{RT}{2\mathcal{F}} \ln \gamma_{Zn^{++}}\gamma_{Cl^-}^2 \qquad [7.35]$$

However, the mean activity coefficient is defined by:

$$\gamma_\pm^3 = \gamma_{Zn^{++}}\gamma_{Cl^-}^2 \qquad [7.36]$$

Relation [7.35] can then be written in the form:

$$E_{\text{abs}} = E^0 - \frac{RT}{2\mathcal{F}} \ln 4C^2 - \frac{3RT}{2\mathcal{F}} \ln \gamma_\pm \qquad [7.37]$$

When we know the standard potential of the cell E^0 and the concentration, we deduce the mean activity coefficient of the ions by measuring E_{abs} and using the following relation, derived from the former:

$$\ln \gamma_\pm = \frac{2\mathcal{F}}{RT}\left(E^0 - E_{\text{abs}} - \frac{1}{3} \ln 4C^2 \right) \qquad [7.38]$$

Note that it is wise to create junctionless cells because, however perfect the junctions may be, they can cause significant errors – even up to 50% in terms of the value of the cell's voltage.

7.7.4. *Influence of complex formation on the redox potential*

We shall now show that the formation of complexes may alter the normal potential of a redox couple. This process is used to stabilize ionic forms – usually cations – whose instability in the aqueous phase prevents us from measuring the redox potentials of the couples in which those ions are involved.

This instability may stem from a number of causes. Here we shall cite two, which are linked to oxidation/reduction:

– the case of cations that are so strongly oxidant that they oxidize the solvent;

– the case of cations undergoing a dismutation.

To illustrate the former case, we shall look at the couple Co^{3+}/ Co^{2+}. The trivalent cobalt is so strongly oxidant that it oxidizes the water:

$$Co^{3+} + 1/2H_2O \rightarrow Co^{2+} + H^+ + 1/2O_2. \qquad [7R.10]$$

Hence, it is not possible to measure the normal redox potential of the couple $e^0_{Co^{3+}/Co^{2+}}$ because the Co^{3+} ions are unstable in water. We can use the hexammine complexes of cobalt $Co(NH_3)_6^{3+}$ and $Co(NH_3)_6^{2+}$. The dissociation constants of these complexes are, respectively:

$$K_D^{(3+)} = 2.4\times10^{-34} \quad [7.39]$$

$$K_D^{(2+)} = 1.2\times10^{-5} \quad [7.40]$$

Consider an ammonia solution containing the same concentration C of those two complexes. In practical terms, we can write:

$$\left[Co(NH_3)_6^{3+}\right] \cong \left[Co(NH_3)_6^{2+}\right] \quad [7.41]$$

and:

$$\frac{\left[Co^{3+}\right]}{\left[Co^{2+}\right]} = \frac{K_D^{(3+)}}{K_D^{(2+)}} = 2\times10^{-29} \quad [7.42]$$

The voltage of the couple is:

$$e_{Co^{3+}/Co^{2+}} = e^0_{Co^{3+}/Co^{2+}} + 0.06\log\frac{\left[Co^{3+}\right]}{\left[Co^{2+}\right]} \quad [7.43]$$

This voltage is impossible to measure in water, because the ratio $\frac{\left[Co^{3+}\right]}{\left[Co^{2+}\right]}$ is practically zero. In the presence of the complex, we calculate:

$$e_{Co^{3+}/Co^{2+}} = e^0_{Co^{3+}/Co^{2+}} + 0.06\log 2\times10^{-29} = 0.12 \quad [7.44]$$

The valence Co^{3+}, which has become less strongly oxidizing, is therefore stabilized, and we can measure the value of $e^0_{Co^{3+}/Co^{2+}}$, which is 1.82 V.

The second case of instability of a degree of oxidation is that of dismutation. For example, it is impossible to stabilize the Cu^+ ion in water, because it dismutes, giving:

$$2Cu^+ = Cu + Cu^{2+} \quad [7R.11]$$

with the constant:

$$\frac{\left[Cu^{2+}\right]}{\left[Cu^{+}\right]^{2}}=10^{6}$$

However, in the presence of chlorides, Cu^{+} yields complexes that are more stable than those deriving from Cu^{2+}. This complex formation leads to a shift of equilibrium [7R.11] to the left, and thus monovalent copper can exist in the form of cupric chloride $CuCl_2^-$.

Hence, we can stabilize unusual degrees of oxidation, such as Ag^{2+}, Ni^{3+}, Cu^{3+}, etc.

7.7.5. *Electrochemical methods for studying complexes*

In section 4.1.3.2, we mentioned the existence of electrochemical methods to study the characteristics of a complex ion: its dissociation constant and its coordination number. We shall now examine those methods, looking at three in particular:

– the potentiometric method;

– the method using a redox electrode;

– the polarographic method.

7.7.5.1. *Studying complex ions by the potentiometric method*

Consider a cation M^{z+} which gives a complex $(ML_n)^{z+}$ with a ligand L. The stability constant is written:

$$K_D=\frac{\left[M^{z+}\right]\left[L\right]^n}{\left[ML_n^{z+}\right]} \qquad [7.45]$$

Remember that we are trying to determine n and K_D.

We prepare a series of solutions whose total concentration of salt is M^{z+} and which contain a variable excess x of the ligand carrier. The solutions are sufficiently dilute for us to be able to treat the activities and concentrations as being the same thing.

Using an indicator electrode for the M^{z+} ions, we measure:

$$pM = -\log\left[M^{z+}\right] \qquad [7.46]$$

To do this, we create a concentration cell with, on one side, the solution prepared above (compartment 1) and, in the second compartment, a solution whose concentration is C but does not contain any ligand carrier ($x = 0$).

The electromotive force of that cell is:

$$E = 0.06\log\frac{\left[M^{z+}\right]_2}{\left[M^{z+}\right]_1} = 0.06\left(\log C + pM\right) \qquad [7.47]$$

By measuring E, we are able to calculate pM.

We shall now examine two cases, depending on whether or not the complex is perfect.

7.7.5.1.1. Case of a perfect complex

Because the complex is stable, we can write:

$$\left[ML_n^{z+}\right] = C \qquad [7.48]$$

and because the ligand carrier is present in excess, in practice, we have:

$$\left[L\right] = x \qquad [7.49]$$

Relation [7.45] is then written:

$$K_D = \frac{\left[M^{z+}\right]x^n}{C} \qquad [7.50]$$

Thus:

$$pM + \log C = -\log K_D + n\log x \qquad [7.51]$$

Thus if, for the various solutions containing introduced quantities of ligand x, we plot the quantity pM + log$C$ as a function of log x (see Figure 6.3(a)), we should obtain a straight line with the slope n and with the ordinate value –log K_D at the origin.

7.7.5.1.2. Case of an imperfect complex

In the case of an imperfect complex, the metal ions are not all complexed, and instead of relation [7.48], we have the expression:

$$\left[\mathrm{ML}_n^{z+}\right] = C - \left[\mathrm{M}^{z+}\right] \quad [7.52]$$

Employing the same logic as before, we find the expression:

$$\mathrm{pM} + \log\left(C - \left[\mathrm{M}^{z+}\right]\right) = -\log K_D - + n \log x \quad [7.53]$$

Thus time (see Figure 7.3(b)), we plot the term $\mathrm{pM} + \log\left(C - \left[\mathrm{M}^{z+}\right]\right)$ as a function of log x. Once again, for the slope and the ordinate at the origin, we obtain n and K_D.

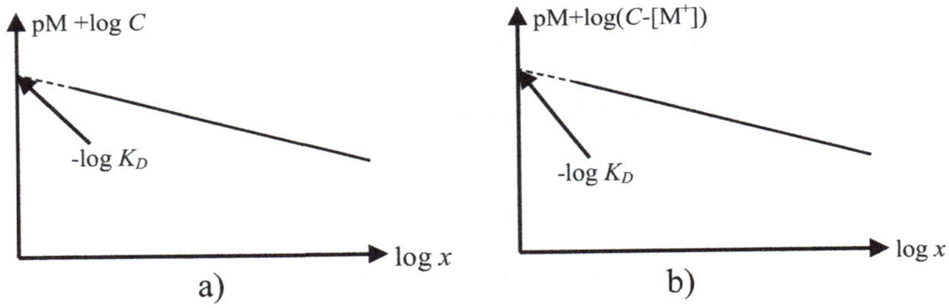

Figure 7.3. *Determination of the characteristics of a complex by potentiometry: a) perfect complex; b) imperfect complex*

NOTE.– It may happen that the curves shown in Figure 7.3(a) and 7.3(b) are not straight lines. This variable slope then implies the existence of several successive complexes of differing compositions. Faced with this result, we need to adopt a different approach, because then the slope at each point of the curve signifies a mean coordination index. We can therefore plot the value of that slope as a function of log x, and examine multiple hypotheses depending on the value taken by that mean value of the coordination index.

7.7.5.2. *Studying complex ions by the use of a redox electrode*

Let us now consider a metallic element liable to yield cations of two different degrees of oxidation, which we denote as Rd and Ox. Suppose that both of these degrees of oxidation, or only one of them, will form complex ions with a ligand L. Those complexes are of the form OxL_n and $RdL_{n'}$. Between the two degrees of oxidation, there is a redox equilibrium which is written:

$$\mathrm{Ox} + \nu_e \mathrm{e} = \mathrm{Rd} \qquad [7R.12]$$

The normal potential of the couple is e_{12}^0.

We create a concentration cell whose first compartment (compartment 1) contains equal concentrations of the two forms Ox and Rd; the second compartment is also filled with the two forms Rd and Ox at the same concentration, but this time, we add a large excess of the complexing agent L, in the amount x which varies from one experiment to the next. The effect of introducing the complexant is to alter the concentrations of the two differently, because they do not behave in the same way towards the complexant.

The potential of the electrode in compartment 1 is:

$$e_1 = e_{11}^0 \qquad [7.54]$$

The potential of the electrode in compartment 2 will be:

$$e_1 = e_{11}^0 + \frac{0.06}{\nu_e}\log\frac{[\mathrm{Ox}]}{[\mathrm{Rd}]} \qquad [7.55]$$

The electromotive force of the cell will therefore be:

$$E = e_1 - e_2 = \frac{0.06}{\nu_e}\log\frac{[\mathrm{Ox}]}{[\mathrm{Rd}]} \qquad [7.56]$$

Thus, by measuring E, we can calculate the concentration of Rd when we know the concentration of Ox, or vice versa.

Initially, we shall only image the existence of one complex – e.g. OxL_n. In such a case, the concentration of Rd in compartment 2, which is not modified by the addition of the complexing agent, is known. We are therefore able to calculate the concentration of Ox in the cell's second compartment – pOx. We are led to use the potentiometric method (see section 7.7.5.1) because we can determine pOx at different values of x.

Now suppose there are two complexes OxL_n and $RdL_{n'}$, whose respective dissociation constants are K_D and K'_D. We suppose that the two complexes formed are sufficiently stable, so the concentration in each of the complexes would be the same and practically equal to the quantities of Rd and of Ox initially introduced. Thus, we write the two expressions of stability:

$$K_D = \frac{[\mathrm{Ox}]\ [\mathrm{L}]^n}{[\mathrm{Ox}]_0} \qquad [7.57]$$

$$K'_D = \frac{[\mathrm{Rd}]\ [\mathrm{L}]^{n'}}{[\mathrm{Rd}]_0} = \frac{[\mathrm{Rd}]\ [\mathrm{L}]^{n'}}{[\mathrm{Ox}]_0} \qquad [7.58]$$

From these two expressions, we deduce:

$$\frac{[\mathrm{Ox}]}{[\mathrm{Rd}]} = \frac{K'_D}{K_D}[\mathrm{L}]^{(n'-n)} \qquad [7.59]$$

When we feed this ratio back into the expression of the potential of the cell (equation [7.56]), we find:

$$E = \frac{0.06}{\nu_e}\log\frac{\mathrm{K_D}}{\mathrm{K'_D}} + (n'-n)\frac{0.06}{\nu_e}\log x \qquad [7.60]$$

If one of the two complexes is known (e.g. n and K_D), then by measuring the above voltage at different values of x, we are able to obtain a straight line, plotting E as a function of log x. The slope of that line can be used to calculate the value of n' and its ordinate at the origin, K'_D, which are the characteristics of the second complex.

7.7.5.2.1. Studying complex ions by polarography

Polarography is a technique whose principle is not based on the thermodynamics of electrochemistry; it involves intensity/potential curves, which indubitably come from the realm of electrochemical kinetics. However, without going into detail on the underlying theory (for which, interested readers can refer to publications in electrochemical kinetics – e.g. J. Besson [BES 84]), we believe it is necessary to touch upon this method, because it involves the study of the thermodynamic properties of a complex ion, and in particular, its dissociation constant.

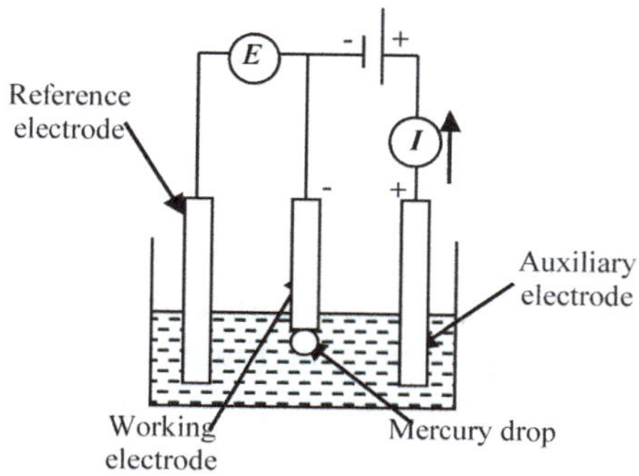

Figure 7.4. *Principle of polarography – diagram of device with three electrodes*

Consider a solution of a cation $M^{\nu+}$ which forms a redox couple that can be written in the form:

$$M^{\nu+} + \nu e = M \qquad [7R.13]$$

That couple is characterized by a redox voltage e_{13}^{0}.

We perform electrolysis of the solution in a device with three electrodes, illustrated by Figure 7.4. That device includes an electrochemical cell composed of the solution of the cation, a mercury hanging electrode called the *working electrode* and an auxiliary electrode. A current generator provides an electrical current of intensity I in the circuit, measured by an amperometric device. A third electrode, known as the *reference electrode*, maintains a potential E on the working electrode. A voltametric device is used to find that voltage. The resistance of the circuit thus constituted with the working electrode is such that the current passing through it is negligible in relation to I.

If the conditions are correct – if the solution has a pH buffer added to it; if the electrolyte contains a spectator salt which is responsible for almost all of the transport of current in the electrolyte; if the process of electrolysis is determined by diffusion in the double layer of the electrode; and if the process is sufficiently quick so that the electrode is practically always at equilibrium – then, if we set a value of the voltage E, we obtain a current I which obeys the following relation, stemming from Nernst's and Faraday's laws:

$$E = e = e_{12}^{0} - \frac{0{,}06}{\nu_e}\log\frac{I}{I_d - I} \qquad [7.61]$$

In this expression, I_d is the *intensity of diffusion*, which depends only on the reduced redox couple on the mercury cathode. The function I deduced from relation [6.61] is represented in Figure 7.5(a). This curve, which is known as the *polarogram* or *polarographic wave*, is plotted by placing the cathodic current I on the ordinate axis and the opposite of the mercury cathode voltage, $-E$, on the abscissa axis. (In choosing $-E$, we obtain a positive value, because the cathode voltage is negative). The curve displays an inflection point for $I = I_d/2$. The abscissa of that point is called the *half-wave potential*. It is a characteristic of the redox couple [7R.13].

The $M^{\nu+}$, when combined with a donor of the ligand L, is likely to form a complex ML_p ,with coordination number n and dissociation constant K_D, by the reaction:

$$M + n\,L = ML_n \qquad [7R.14]$$

Using the theory of electrolysis, performed in the conditions we saw earlier, it is possible to show that the half-wave potential of the couple [7R.14] is shifted along the potential axis, to the right, by a distance $\Delta E_{1/2}$ such that:

$$-\Delta E_{1/2} = -E_p - E_0 = -\frac{0.06}{\nu_e}\log K_D + \frac{0.06}{\nu_e}n\log[L] \qquad [7.62]$$

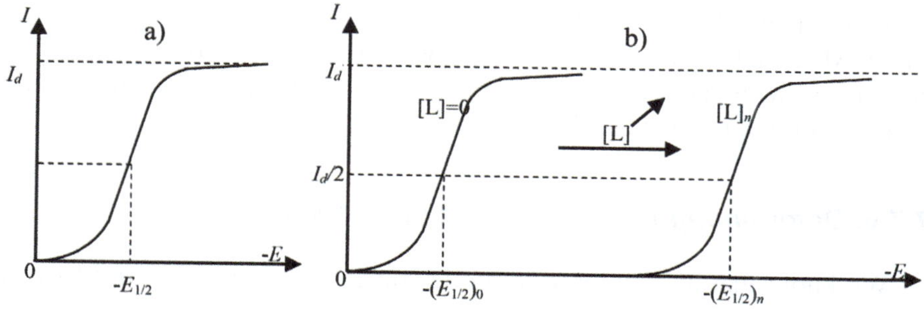

Figure 7.5. *Polarographic waves: a) with a single cation; b) cation alone and complexed*

Figure 7.5(b) shows the shift of the polarization wave caused by the formation of the complex with coordination number n, obtained by adding in the ligand L.

If, using relation [7.62], we represent the value $-\Delta E_{1/2}$ as a function of the decimal logarithm of the concentration of ligand introduced (see Figure 7.6(a)), throughout the domain of predominance of the complex ML_n, we obtain a straight line whose slope is $\frac{0.06}{\nu_e}n$ and whose ordinate at the origin is $-\frac{0.06}{\nu_e}\log K_D$ (generally speaking, K_D is much lower than 1, and therefore that ordinate value is positive). Hence, by measuring that slope and that ordinate at the origin, we are able to calculate our two characteristics of the complex: n and K_D.

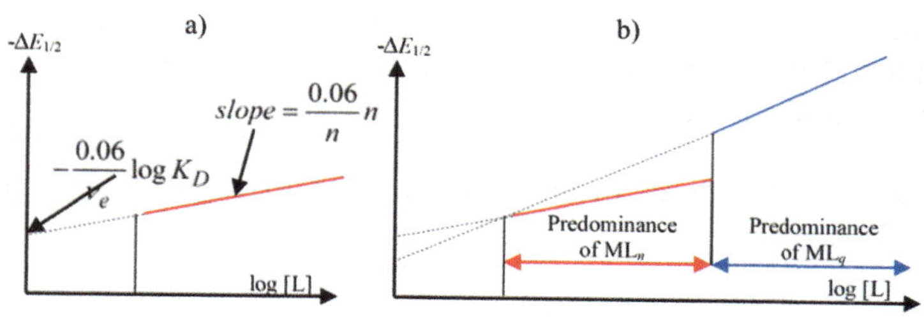

Figure 7.6. *Determination of the dissociation constant and the coordination index by polarography: a) for a complex; b) for two successive complexes. For a color version of the figure, see www.iste.co.uk/soustelle/equilibria.zip*

If the ligand L is liable to form several successive complexes with the cation $M^{\nu+}$, such a line segment will appear in Figure 7.6(b) for each of the complexes with index n, q, ($q > n$)… Each segment corresponds to the domain of predominance of the corresponding form.

7.7.6. *Determining the ion product of a solvent*

We know that an amphiprotic protic solvent SH is characterized by an equilibrium constant which is its ion product: K_i, such that:

$$K_i = \left|SH_2^+\right| \left|S^-\right| \qquad [7.63]$$

We want to determine that constant on the basis of a potentiometric measurement.

In order to do so, we set up the following concentration cell, formed of two hydrogen electrodes immersed in an electrolyte which, in one of the compartments (compartment 2), contains the solvent under examination and a strong base completely dissociated in that solvent, and in compartment 1, a completely-dissociated strong acid and our solvent.

	(2)		(1)	
H_2 (1 bar) |	CSH (0.01)	||	XH (0.01)	| H_2 (1 bar)

To measure the ion product of water, we might use, say, potassium hydroxide in compartment 2 and hydrochloric acid in compartment 1.

To measure the ion product of ethanolamine, the strong base in compartment 2 would be ethanolamine sodium salt NH_2-CH_2-CH_2Ona, and the strong acid would be an ethanolammonium salt.

The electromotive force of this cell, which is a concentration cell, would be:

$$E = e_2 - e_1 = \frac{RT}{\mathcal{F}} \log \frac{\left[H_3O^+\right]_2}{\left[H_3O^+\right]_1} \qquad [7.64]$$

In view of the fact that only low concentrations are being used, we treat the activities and concentrations as being one and the same. In these conditions, we have:

$$\left[H_3O^+\right]_2 = 0.01 \qquad [7.65]$$

and

$$\left[H_3O^+\right]_1 = \frac{K_i}{\left[S^-\right]_1} = \frac{K_i}{0.01} \qquad [7.66]$$

By substituting these values back into relation [7.64], we can deduce:

$$E = \frac{RT}{\mathcal{F}} \log \frac{(0.01)^2}{K_i} \qquad [7.67]$$

Thus, by measuring the electromotive force, we can calculate the ion product of the solvent.

7.7.7. Determining a solubility product

A solubility product of a salt $A_\nu M$ is the equilibrium constant of precipitation, in accordance with:

$$\nu A^- + M^+ = A_\nu M \text{ (solid)} \qquad [7R.15]$$

This product is expressed by the relation:

$$K_S = \left[A^-\right]^\nu \left[M^{\nu+}\right] \qquad [7.68]$$

The activities are replaced by the concentrations, in view of the very poor solubility of the salt $A_\nu M$. The dissolved concentrations are very low indeed.

To obtain this equilibrium constant by potentiometric measurement, we merely need to think of a cell which operates by the reaction [7R.15].

Consider a second-species metal electrode (see section 6.3.2.2). Remember that this is an electrode in which the metal is at equilibrium with a second solid phase containing the same element as the metal and a solution containing an ion of an element from the second solid phase.

We then constitute a cell using that electrode, a solution containing a electrolyte with the same anion as the product $A_{\nu}M$ and an indicator electrode for the cations C whose potential e_C is known.

M , AM solid	$A_{\nu}M$ (c)	$A_{\nu}C$	C

This cell operates by reaction [7R .15]. Its electromotive force is:

$$E = e_C^0 - e_M^0 - \frac{0.06}{\nu}\log\left[A^-\right]^{\nu}\left[M^{\nu+}\right] = e_C^0 - e_M^0 - \frac{0.06}{\nu}\log K_S \qquad [7.69]$$

We can see that measuring the electromotive force of our cell will help to calculate the solubility product K_s if we know the potentials e_C^0 and e_M^0.

7.7.8. Determining the enthalpies, entropies and Gibbs energies of reactions

We can determine the main functions for the reaction – the enthalpies, entropies and Gibbs energies – on the basis of the measured values of the emf of cells.

We merely need to set up a cell whose operational reaction is the reaction under study, R. We then measure the electromotive force of that cell at two temperatures that are not too far apart, T_1 and T_2, and let E_1 and E_2 be the measure values.

We can calculate the cell temperature coefficient (see section 7.4):

$$\frac{\mathrm{d}E}{\mathrm{d}T} = \frac{E_2 - E_1}{T_2 - T_1} \qquad [7.70]$$

From relation [7.8], we deduce the entropy of the reaction:

$$\Delta_R S = -\nu_e \mathfrak{F} \frac{\mathrm{d}E}{\mathrm{d}T} \qquad [7.71]$$

From relation [7.7], we calculate the enthalpy of the reaction:

$$\Delta_R H = \nu_e \mathfrak{F} E_2 + T \frac{dE}{dT} \quad [7.72]$$

The Gibbs energy is obviously:

$$\Delta_R G = \nu_e \mathfrak{F} E_2 \quad [7.73]$$

The temperature coefficients of the cells are around 10^{-4} volts/degree.

7.7.9. *Determining the standard Gibbs energies of the ions*

The standard Gibbs energy of a cation is the standard Gibbs energy of the reaction:

$$M = M^{+\nu} + \nu e \quad [7R.16]$$

In fact, as it is impossible to find a value for an ion on its own, the standard Gibbs energy is, in fact, a value relative to that of the reaction:

$$\frac{1}{2} H_2 = H^+ + e \quad [7R.17]$$

For this reaction, we accept that the standard Gibbs energy is 0 at all temperatures.

Hence, the relative Gibbs energy is, in fact, the standard Gibbs energy of the reaction:

$$\Delta G_M^0 = \Delta G_{16}^0 - \Delta G_{17}^0 \quad [7.74]$$

To measure that standard Gibbs energy, we simply need to determine the standard potential of the electrode operating by way of reaction [7R.16] (see section 7.7.1), by creating a cell coupling that electrode with a hydrogen reference electrode.

The method examined above is not always easy to apply in the case of anions, because there are not many specific anion electrodes. We then use the indirect method, calculating on the basis of the solubility products.

7.7.10. *Determining the standard entropies of the ions*

We use relation [7.8], applied to the standard entropy:

$$\Delta S_M^0 = -\nu_e \mathfrak{F} \frac{\mathrm{d}e^0}{\mathrm{d}T} \quad [7.75]$$

The meaning of ΔS_M^0 is:

$$\Delta S_M^0 = \left(S_{M^+}^0 - S_M\right) - \left(S_{H^+}^0 - S_{H_2}\right) = S_{M^+}^0 - S_{H^+}^0 \quad [7.76]$$

We are still dealing with a value relative to hydrogen, for which we established:

$$S_{H^+}^0 = 0 \quad [7.77]$$

If we know the entropies of the atoms $S_{\mathrm{H}_2}^0$ and S_M^0, by measuring the value of e^0, we are able to calculate:

$$S_{\mathrm{M}^+}^0 = -\nu_e F \frac{\mathrm{d}e^0}{\mathrm{d}T} + S_{\mathrm{M}}^0 - S_{\mathrm{H}_2}^0 = -\nu_e F \frac{\mathrm{d}e^0}{\mathrm{d}T} + S_{\mathrm{M}}^0 \quad [7.78]$$

For anions, we encounter the same difficulty as in the case of the standard Gibbs energy (see section 7.7.10).

7.7.11. *Measuring the activity of a component of a non-ionic conductive solution (metal solution)*

If we consider an electrode reaction which would be written:

$$\sum_i \mathrm{v_i M_i} + \mathrm{v_e e} = 0 \quad [7R.18]$$

the potential is given by Nernst's law.

$$e = e^0 + \frac{\mathrm{R}T}{\nu_e \mathfrak{F}} \ln \prod \left|\mathrm{M}_i\right|^{\nu_i} \quad [7.79]$$

This expression involves the activities of all the species M_i involved in equilibrium [7R.18].

In general, the electrode is composed of a pure metal, whose activity is equal to 1. However, if the electrode is made of an alloy forming a solid- or liquid solution, then relation [7.79] will include the activity of the active metal in the allow. However, relation [7.79] contains various other activities that are usually not known. Let us examine the problem using a concrete example.

For instance, we can envisage measuring the activity of lead in an amalgam of mercury and lead, which has the molar fraction x_1 of lead. The electrode is made up of the alloy under study and a lead salt in which the activity of the lead is a_2, so for example:

$$\text{Pb (amalgam } (x_1)\text{)} \mid \text{Pb}(\text{CH}_3\text{COO})_2\,(a_2)$$

The potential of that electrode is written:

$$e = e^0 + \frac{RT}{2\mathcal{F}}\frac{a_2}{a_1} \qquad [7.80]$$

The activity a_1 of the lead cannot be deduced from the measured value of the potential of that electrode, unless the activity of the lead in the solution is known.

To circumvent this problem, we construct a cell, similar to concentration cells, but in this case, the dissymmetry will be created by a difference in concentration between the two electrode alloys, rather than by two ionic compartments.

We place our electrode in a cell where the other electrode is the same amalgam but at a different molar fraction of lead x_3. The two electrodes are immersed in a solution of a lead salt – acetate, for example. Thus, the following electrode is created:

$$\text{Pb(amalgam}(x_3)\text{)} \mid \text{Pb}(\text{CH}_3\text{COO})\text{CH}_3\text{COOH} \mid \text{Pb(amalgam}(x_1)\text{)}$$

For both two electrodes, the reference state of the lead is the same, so the cell's electromotive force is written:

$$E = \frac{\mathrm{R}T}{2\mathfrak{F}} \ln \frac{a_1}{a_3} \tag{7.81}$$

If we separate the molar fraction and the activity of lead in the known amalgam, this expression can be rewritten in the form:

$$\frac{2\mathfrak{F}E}{\mathrm{R}T} + \ln x_1 = \ln a_3 - \ln \gamma_1 \tag{7.82}$$

As the molar fraction of lead in our solution x_1 tends toward zero, the activity coefficient $\gamma_1^{(II)}$ tends toward 1. Hence, we can write:

$$\lim_{x_1 \to 0} \left(\frac{2\mathfrak{F}E}{\mathrm{R}T} + \ln x_1 \right) = \ln a_3^{(II)} \tag{7.83}$$

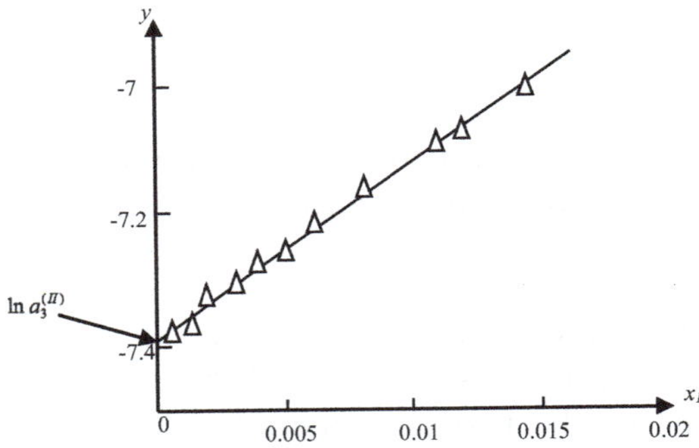

Figure 7.7. *Obtaining the activity of an alloy (from [ROS 85])*

In order to use this latter relation, we shall therefore measure the electromotive force of the cell at different values of the molar fraction x_1, and plot the curve showing the value $y = \frac{2\mathfrak{F}E}{\mathrm{R}T} + \ln x_1$ as a function of the molar fraction x_1. Figure 7.7 shows an example of such a curve.

By extrapolating that curve to the abscissa origin, we are able to find the value of $\ln a_2^{(II)}$, which is the activity of lead in the second compartment. Armed with this value, it becomes easy to calculate the sought activity coefficient for each concentration, using relation [7.82], in the form:

$$\ln \gamma_1^{(II)} = -\frac{2\mathfrak{F}E}{RT} - \ln x_1 + \ln a_3^{(II)} \qquad [7.84]$$

The method is particularly elegant, as it does not require the use of a junction.

7.7.12. *Measuring the activity coefficient of transfer of a strong electrolyte*

We saw in section 2.6.2 that it is only possible to measure the mean activity coefficient of transfer of a strong electrolyte between two solvents, and we described the method pertaining to solubility. It is also feasible, in order to measure that mean coefficient, to employ potentiometry.

For example, to measure the mean activity coefficient of transfer of HCl, we construct the cell shown below, which has a hydrogen electrode in normal conditions, a solution of hydrochloric acid with activity equal to 1 in one of the two solvents under study, and a silver reference electrode.

Pt; H_2 (P = 1 bar)	$\lvert HCl \rvert = 1 + \text{solvent}$	$\lvert AgCl \rvert$solid; Ag (reference electrode)

We measure the electromotive force of that cell, which is the standard electromotive force of HCl in that first solvent S_1 – i.e. $E^0_{S_1}$. We do the same thing with the second solvent S_2 and we obtain the standard potential of HCl in that second solvent, $E^0_{S_2}$.

The difference between the two normal potentials is such that we have:

$$\mathfrak{F}\left(E^0_{S_1} - E^0_{S_2}\right) = \Delta_{S_1 \to S_2} G^0_{HCl} = 4.6RT \log {}^{(S_1)}\Gamma^{(S_2)}_{\pm HCl} \qquad [7.85]$$

We can see that by knowing the two standard potentials, we are able to calculate the mean activity coefficient of transfer of HCl between the two solvents S_1 and S_2.

Similarly, we can replace the hydrogen electrode with an indicator electrode for the ions M^{z+} and HCl with the chloride of that metal, in each of the two solvents, we can then measure the mean activity coefficient of transfer of the salt MCl_z between the two solvents.

7.7.13. *Evaluating the individual activity coefficient of transport for an ion*

As it is impossible to measure the transfer activity coefficient of an ion between two solvents, we have seen that the introduction of *extra-thermodynamic hypotheses* helps to evaluate that coefficient.

7.7.13.1. *Pleskow's method*

We saw earlier that Pleskow advocated the use of the rubidium or cesium ions (see section 2.6.3.1). An electrochemical method, using the couples Rb/Rb+ or Cs/Cs+, can also be used.

Consider the cell illustrated below, composed of a rubidium electrode and a silver electrode:

Rb (amalgam)	$\lvert \mathrm{RbCl} \rvert = 1 + \text{solvent}$	AgCl solid; Ag

This cell can be used to measure the standard chemical potential of rubidium chloride in a solvent. The measurement is repeated on both solvents S_1 and S_2.

However, we can accept that for the rubidium ion, we have:

$$\Delta_{Tr} G^0_{\mathrm{Rb}^+} = 0 \qquad [7.86]$$

Thus, the difference between the standard Gibbs energies of the two measurements of the electromotive force will give an estimation of the Gibbs energy of transfer of the chloride ion.

Similarly, the cell formed of the same rubidium electrode as above with a metal electrode M with a solution of ions M^{z+} in the same solvent, such as the following cell:

$$\text{Rb (amalgam)} \;\Big|\; \left|Rb^{+}\right| = 1 + \text{solvent} \;\Big|\; \left|M^{z+}\right| = 1 + \text{solvant} \;\Big|\; \text{M}$$

can be used to evaluate the activity coefficient of transfer of the cation M^{z+}.

NOTE.– The cell presented above displays a junction electromotive force, but as that junction is created with the same solvent on both sides, the corresponding junction potential is only slight (see section 6.4.2).

7.7.13.2. *Strehlow's method*

We saw (in section 2.6.3.2) that Strehlow advocated an electrochemical method using the ferrocene/ferricinium couple to determine the coefficient of transfer activation of an ion, based on the equality:

$$\Delta_{Tr}G_{total} = \Delta_{Tr}G_{M^{+}} + \left(\Delta_{Tr}G_{\text{ferrocene}} - \Delta_{Tr}G_{\text{ferricinium}}\right) \qquad [7.87]$$

According to Strehlow's hypothesis, the term in parentheses in this expression would have a value of 0.

In order to measure, we set up the following cell:

$$\text{Rb (amalgam)} \;\Big|\; \begin{matrix}\text{Ferrocene} + \text{ferricinium} \\ + \text{solvent}\end{matrix} \;\Big|\; \left|M^{z+}\right| = 1 + \text{solvent} \;\Big|\; \text{M}$$

This cell uses a metal electrode with the cation dissolved in a solvent and an electrode working with the couple ferrocene/ferricinium, dissolved in the same solvent.

We then work in the same way as for the previous method, measuring the cell's potential using each of the two solvents successively, and the measured potential difference gives us the Gibbs energy of transfer between the two solvents, by the relation:

$$\mathfrak{F}\Delta E^{0} = \Delta_{Tr}G_{M^{+}} \qquad [7.88]$$

Thus, we are able to evaluate the activity coefficient of transfer of the M^{z+} ion.

8

Potential/Acidity Diagrams

In the aqueous phase, potential/acidity diagrams are potential/pH diagrams (also known as *e*/pH diagrams), or Pourbaix diagrams. They are thermodynamic diagrams that are very widely used in the chemistry of aqueous solutions. A diagram includes all the potential/pH curves (as shown by Figure 6.1), pertaining to the various redox reactions which involve the compounds deriving from the same element. We also add all the curves relating to the acid–base equilibria in which those compounds play a part. These diagrams are quite similar to Gibbs energy/concentration diagrams, as shown by relation [6.45], and the definition of the pH. Just like *G/C* diagrams, they define domains of predominance for the species in solution, and domains of stability for species in the solid phase.

8.1. Conventions

When plotting these diagrams, we adopt a certain number of conventions, which must always be borne in mind when using them – conventions which can be modified to obtain other diagrams than those appearing in the existing body of literature.

8.1.1. *Plotting conventions*

We agree that the solvent used is water. The solutions are usually fairly dilute – typically 10^{-4} to 10^{-6} moles/liter, so we can treat the activity and the concentration as being one and the same thing. The reference states are the solutions at the concentration of 1 mol/liter, and the activity of the species in the pure solid phase is equal to 1. The gas pressure is set at 1 bar. The reference potential is that of the normal hydrogen electrode (NHE), the only solids in question are oxides and hydroxides, with the exclusion of salts. Finally, the temperature chosen is 25°C, which enables us to use the value 0.06 for the factor 2.3 R$T/\mathfrak{F}$.

The term "plot concentration" denotes the total atomic concentration of an element. For example, the diagram for chlorine compounds shows that element in a variety of forms: the chloride ion, the hypochlorite ion, hypochlorous acid, chlorine dissolved in water, and so on. The plot concentration of chlorine is defined by the sum:

$$C = \left[\mathrm{Cl}^-\right] + \left[\mathrm{ClO}^-\right] + [\mathrm{ClOH}] + [\mathrm{Cl}_2] \quad [8.1]$$

As noted earlier, this concentration in generally low.

8.1.2. *Boundary equations*

A boundary is a line in the diagram which separates two domains of predominance or stability of two forms of the same species. It is characterized by a chemical or electrochemical equilibrium, and therefore an equation *f*(*e*, pH) = 0.

Consider a general equation linking two species A_1 and A_2, containing the same element A, written as:

$$a_1\mathrm{A}_1 + h\mathrm{H}^+ + \nu_e \mathrm{e}^- = a_2\mathrm{A}_2 \quad [8R.1]$$

8.1.2.1. *Acid–base equilibria*

If, in equilibrium [8R.1], we have $\nu_e = 0$, the previous equilibrium is an acid–base equilibrium: A_1 is the basic form and A_2 the acidic form of the couple. Three cases may be encountered, then:

– if both species A_1 and A_2 are in solution, the law of mass action applied to the equilibrium will yield the following, where K_A denotes the acidity constant of A_2:

$$\mathrm{pH} = \frac{1}{h}\mathrm{p}K_A + \frac{1}{h}\log\frac{[\mathrm{A}_1]^{a_1}}{[\mathrm{A}_2]^{a_2}} \quad [8.2]$$

It is helpful to plot the diagram for the following concentrations, known as line concentrations:

$$a_1[A_1] = a_2[A_2] = \frac{C}{2} \qquad [8.3]$$

The corresponding boundary, then, will be a vertical line. To its right (low pH) we have the domain of predominance of the acidic form A_2, and to the left, that of the basic form A_1. This line is sometimes called the equi-concentration line. The transition pH is the pH of equi-concentration of the two forms A_1 and A_2. We can see from relation [8.2] that this pH depends on the concentration C, unless $a_1 = a_2$.

The same relation shows that as we move away a little from the equi-concentration pH, one of the forms very quickly becomes massively predominant.

– if one of the two forms A_1 or A_2 has low solubility, its activity is equal to 1, and the concentration of the soluble form (A_1, for example) is:

$$[A_1] = \frac{C}{a_1} \qquad [8.4]$$

Thus, the equi-concentration pH is given by:

$$\mathrm{pH} = \frac{1}{h}\mathrm{p}K_A + \frac{a_2}{h}\log a_1 - \frac{a_2}{h}\log C \qquad [8.5]$$

The representative curve, again, is a vertical line.

– if one of the acidic or basic forms is a gas, the dissolved form will be at concentration C and the gas pressure of 1 bar. The equation for the line, which is always vertical, will then be as follows, if A_1 is gaseous, for instance:

$$\mathrm{pH} = \frac{1}{h}\mathrm{p}K_A - \frac{a_2}{h}\log C \qquad [8.6]$$

Note that if one of the forms A_1 or A_2 is solid or gaseous, the other must necessarily be in solution, because it has to be an ion in order to preserve the electrical neutrality of the solution in the presence of H^+ ions.

8.1.2.2. *Redox couples of non-oxygenated species*

If, in equilibrium [8R.1], we have $h = 0$, the previous equilibrium represents a redox couple wherein A_1 is the oxidant and A_2 the reductant. Both forms are non-oxygenated. The line in the diagram will be a horizontal straight line.

We use the same plot concentrations as we did for acid–base couples. The corresponding voltage is the equi-activity voltage (see section 6.2.5), and in this case, will be the equi-concentration voltage, written as e_0, and thus we would have:

– if both forms are soluble:

$$e = e_0 = e^0 + \frac{0.06}{\nu_e}\log\frac{(2a_1)^{a_2}}{(2a_2)^{a_1}} + \frac{(a_2 - a_2)0.06}{\nu_e}\log C \qquad [8.7]$$

– if one of the forms is solid:

$$e = e^0 - \frac{0.06}{\nu_e}\log a_2 + \frac{0.06a_1}{\nu_e}\log C \qquad [8.8]$$

– if one of the forms is gaseous at the pressure of 1 bar, we again observe relation [8.8] to be true.

8.1.2.3. *Redox couples involving oxygenated species*

If, in equilibrium [8R.1], we simultaneously have $\nu_e \neq 0$ and $h \neq 0$, the previous equilibrium is still a redox equilibrium, but this time between two forms, at least one of which is an oxygenated form of A.

We again use the same plot concentrations as for acid–base couples. The corresponding voltage (see section 6.2.5) here will be the equi-concentration voltage e_0. The curve will be a sloping straight line:

– if both forms are soluble, Nernst's law gives us:

$$e = e_0 = e^0 - \frac{0.06h}{\nu_e}\text{pH} + \frac{0.06}{\nu_e}\log\frac{(2a_1)^{a_2}}{(2a_2)^{a_1}} + \frac{(a_2 - a_2)0.06}{\nu_e}\log C \qquad [8.9]$$

– if one of the forms is solid (say A_2), we directly obtain:

$$e = e^0 - \frac{0.06h}{\nu_e}\text{pH} - \frac{0.06}{\nu_e}\log a_2 + \frac{0.06a_1}{\nu_e}\log C \quad [8.10]$$

– if both forms are solid, Nernst's relation becomes:

$$e = e^0 - \frac{0.06h}{\nu_e}\text{pH} \quad [8.11]$$

– if one of the forms is gaseous, we use the concentration C for the soluble form, and the gaseous form is taken to be at the pressure of 1 bar;

– if both forms are gaseous, we need to specify whether we are defining e_0 for identical partial pressures of 1 bar or one of the forms dissolved (Henry's law) and the other at the pressure of 1 bar, or both of them in dissolved form at equal concentrations. The choice is made on the basis of the solubility of the gases.

8.2. Intersections of lines in the diagram

Different lines in the diagram intersect with one another – usually at triple points, where three different lines meet, and those intersections display certain properties. Thus, consider a triple point T, which is the point at which the three half-lines O_1, O_2 and O_3 meet, separating three domains of stability (or predominance) of three compounds A_1, A_2 and A_3 containing the same element A.

8.2.1. *Relative disposition of the lines in the vicinity of a triple point*

Consider three reactions [8R.2], [8R.3] and [8R.4], which are similar to [8R.1]:

$$a_1A_1 + h_1H_3O^+ + \nu_1e^- = a_2A_2 \quad [8R.2]$$

$$a'_2\,A_2 + h_2H_3O^+ + \nu_2e^- = a_3A_3 \quad [8R.3]$$

$$a'_1\,A_1 + h_3H_3O^+ + \nu_3e^- = a'_3\,A_3 \quad [8R.4]$$

They are represented by three lines which come together at the point T. The slope of each of these lines, according to relation [8.9], is $-0.06h_i/\nu_i$.

If the three lines meet, than one reaction is necessarily a linear combination of the other two, and as those reactions satisfy the principle of electro-neutrality, we can show that if we have:

$$-\frac{h_1}{\nu_1} < -\frac{h_2}{\nu_2} \qquad [8.12]$$

we necessarily must have:

$$-0.06\frac{h_1}{\nu_1} < -0.06\frac{h_3}{\nu_3} < -0.06\frac{h_2}{\nu_2} \qquad [8.13]$$

Hence, the slope of one line will necessarily lie between the slopes of the other two. From this, we conclude that the prolongation of each of the lines ending at point T lies within the angle less than π formed by the other two lines.

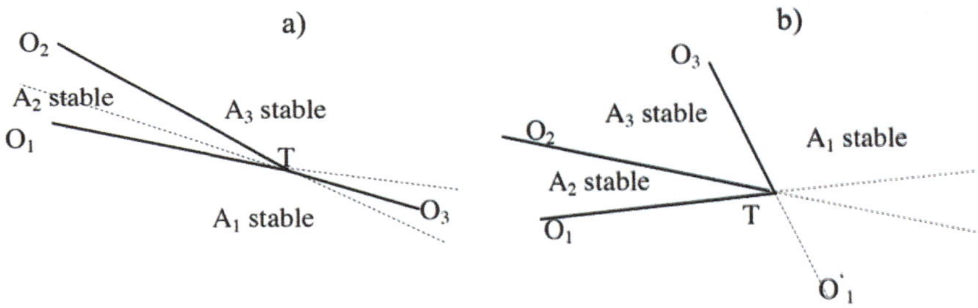

Figure 8.1. *Relative positions of the lines in the vicinity of a triple point*

This means that the relative position of the three lines respects Figure 8.1(a), whereas the scenario shown in Figure 8.1(b) is not acceptable.

8.2.2. *Shape of equi-concentration lines in the vicinity of a triple point*

Up until now, we have defined the different lines in the diagram as being straight-line segments. In fact, small corrections need to be made in certain cases, due to the fact that the line concentrations of the different lines ending at a triple point are not always identical. We distinguish between three cases depending on the physical state (pure solid or species in solution) of the three stable components around the triple point.

8.2.2.1. *Case of a single soluble species*

Suppose that the compound A_1 is in solution, whilst A_2 and A_3 are solids. In Figure 8.2(a), we can see that the conditions of equal concentration and activity imposed on each one of the lines are compatible at the triple point, where we have $|A_2| = |A_3| = 1$ and $[A_1] = C$.

8.2.2.2. *Case of two soluble species*

Now suppose that it is species A_2 and A_3 which are in solution, whilst A_1 is solid. Figure 8.2(b) shows that the conditions of equal concentration of the three straight lines are not compatible at their point of intersection T. The concurrence point where those conditions would be compatible would be a point T', which lies on a curve having lines O_1 and O_2 as its asymptote. Along that curve, the concentrations of A_1 and A_2 are no longer constant, which explains why it is no longer a straight line.

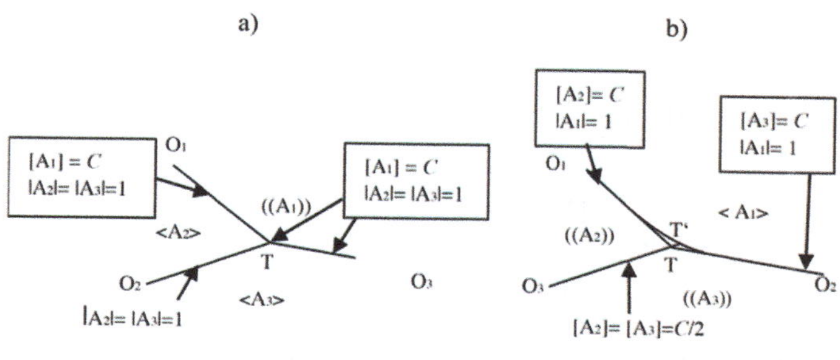

Figure 8.2. *Shapes of the lines in the vicinity of the triple point: a) case of a single soluble species; b) case of two soluble species*

We shall now establish the equation for that new curve. In order to do so, we simply need to write that, along that line, the total concentration of element A in forms A_2 and A_3 is constant.

Suppose, for the sake of simplicity of the calculations, that the two compounds A_2 and A_3 contain only one atom of element A (otherwise, the next equation would be modified but the method of calculation would remain the same). Because that element must be conserved, we are able to write:

$$[A_2]+[A_3]=C \qquad [8.14]$$

Let us write the reactions of the couples A_1 and A_2 and of A_1 and A_3 in the form:

$$A_1 + hH_3O^+ + qH_2O + \nu_e e = A_2 \qquad [8R.5]$$

$$A_1 + h'H_3O^+ + q'H_2O + \nu_e' e = A_3 \qquad [8R.6]$$

By the difference between them, we deduce:

$$A_2 + (h'-h)H_3O^+ + (q'-q)H_2O + (\nu_e'-\nu)e = A_3 \qquad [8R.7]$$

The equation of the curve $T'O_1$, therefore, takes the form:

$$e = e^0_{A_1A_2} + \frac{0.06}{\nu_e}\log\frac{\left[H_3O^+\right]^h}{[A_2]} \qquad [8.15]$$

and the equation for the curve $T'O_2$ is of the form:

$$e = e^0_{A_2A_3} + \frac{0.06}{\nu_e'}\log\frac{\left[H_3O^+\right]^{h'}}{[A_3]} \qquad [8.16]$$

By drawing the concentrations $[A_2]$ and $[A_3]$ respectively from equations [8.15] and [8.16] and feeding those values back into equation [8.14], we obtain the equation for the curve $O_1T'O_2$:

$$\left[H_3O^+\right]^h \exp\left[-\frac{\nu_e}{0.06}\left(e-e^0_{A_1A_2}\right)\right] + \left[H_3O^+\right]^{h'} \exp\left[-\frac{\nu_e'}{0.06}\left(e-e^0_{A_1A_3}\right)\right] = C \qquad [8.17]$$

It is possible to solve this equation analytically if each of the two terms h and h' is less than or equal to 2. If this is not the case, then we need to solve it by the numerical method. Nonetheless, in all cases, a simple geometric construction can be used to obtain as many points on the curve as we want.

Let us begin by constructing point T’ (Figure 8.3). Because it is on the line of equi-concentration of equilibrium A_2A_3, that point corresponds to the concentrations:

$$[A_2]=[A_3]=\frac{C}{2} \tag{8.18}$$

The corresponding voltage is:

$$e_{T'}=e^0_{A_1A_2}+\frac{0.06}{\nu_e}\log\frac{\left[H_3O^+\right]^h}{C}+\frac{0.06}{\nu_e}\log\frac{2}{C} \tag{8.19}$$

The equation of the line for equilibrium A_1A_2, which corresponds to a concentration of A_2 equal to C, is:

$$e_T=e^0_{A_1A_2}+\frac{0.06}{\nu_e}\log\frac{\left[H_3O^+\right]^h}{C} \tag{8.20}$$

Thus, the point T’ is (as shown by Figure 8.3) located on a straight line $T'O'_1$ parallel to TO_1 but shifted toward the domain of A_1 by a value:

$$\Delta e=\frac{0.06}{\nu_e}\log 2=\frac{0.0018}{\nu_e} \tag{8.21}$$

The intersection of that line with that which corresponds to the equilibrium A_2A_3 gives us T’.

More generally, we can determine the position of any given point T’’ on the curve which corresponds to a value of the ratio of concentrations $[A_2]/[A_3]$. For example, let us construct the point T’’, which corresponds to a value of 10 for that ratio – in other words, in view of equation [8.14], for the concentrations:

$$[A_2]=\frac{10C}{11} \tag{8.22a}$$

and $$[A_3]=\frac{C}{11} \tag{8.22b}$$

That point corresponds, for the equilibrium A_2A_3, to a voltage:

$$e = e^0_{A_2A3} - \frac{0.06}{\nu_e - \nu_e^{'}} \log\left[H_3O^+\right]^{(h'-h)} \quad [8.23]$$

The point T'' is therefore located on a line parallel to TO_3, shifted by $\frac{0.06}{\nu_e - \nu_e^{'}}$ volts parallel to the voltage axis. Additionally, in relation to the equilibrium A_1A_2, its voltage is:

$$e = e^0_{A_1A2} + \frac{0.06}{\nu_e} \log\frac{\left[H_3O^+\right]^h}{C} + \frac{0.06}{\nu_e} \log\frac{11}{10} \quad [8.24]$$

The point T'', therefore, is on a parallel to TO_1, shifted towards the domain of A_1 by a value:

$$\Delta e = \frac{0.06}{\nu_e} \log\frac{11}{10} = \frac{0.0025}{\nu_e} \quad [8.25]$$

Figure 8.3 has been plotted on a greatly stretched scale of voltages and pH levels for particular values of the coefficients h, h', ν_e and $\nu_e^{'}$. We can see that the true curve is distinguishable from the asymptotic lines O_1T and O_2T only by a very slight rounding in the vicinity of the point T. As the error is minimal and very localized, we content ourselves with the representation by asymptotic lines plotted at constant concentration.

8.2.2.3. *Case of three soluble species*

In this case (illustrated by Figure 8.4), the equi-concentration curves are half-lines which correspond to inconstant activities, as can be seen in the figure, where we have chosen components each containing one atom of the component A. Indeed, far from the triple point, we have – on the curve TO_1, for example:

$$[A_2] = [A_2] = \frac{C}{2} \quad [8.26]$$

and as we approach the triple point, the content of A_3 increases, so that at T we have:

$$[A_1]=[A_2]=[A_3]=\frac{C}{3} \qquad [8.27]$$

We can see that, in this case, the half-lines are the real curves.

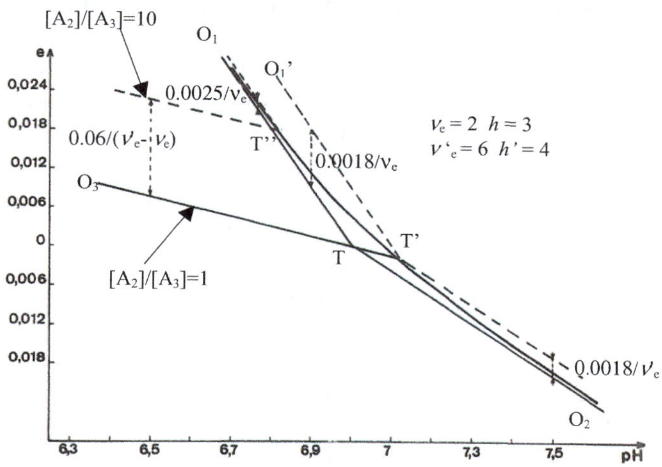

Figure 8.3. *Construction of the line in the vicinity of a triple point with two soluble species*

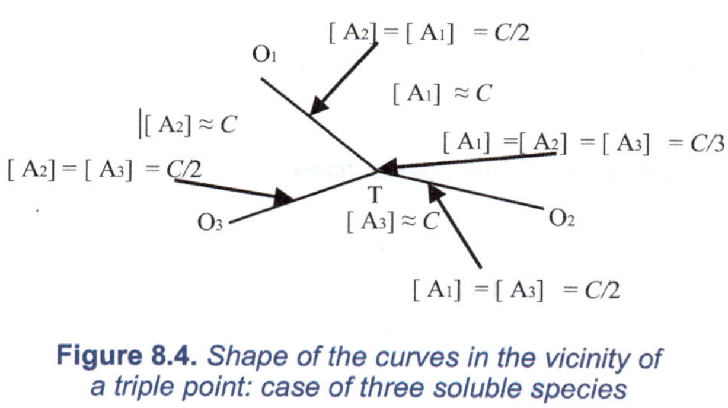

Figure 8.4. *Shape of the curves in the vicinity of a triple point: case of three soluble species*

8.3. Plotting a diagram: example of copper

We have chosen to plot the diagram of copper, to exemplify and fine-tune our methodology, which must be highly rigorous, and during the course of which we must make various choices, such as the existence of oxides and hydrates and the involvement of the phenomenon of dismutation.

8.3.1. *Step 1: list of species and thermodynamic data*

The first step in constructing a diagram is to list all of the different entities to be taken into account. Obviously, species which are not taken into account will show no domain of stability or predominance. This list will be supplemented by the standard chemical potential of formation of each of these compounds, which can be read from the tables. Table 8.1 shows all of these data in the case of copper-based compounds.

Phases	Compounds	μ^0
Dissolved substance	Cu^+	50,160.0 J
	Cu^{2+}	65,915.4 J
	$HCuO_2^-$	-256,735.2 J
	CuO_2^{2-}	-181,830.0 J
	H_2O	-236,964.2 J
Solid substance	Cu_2O	-146,216.4 J
	CuO	-127,072.0 J
	$Cu(OH)_2$	-356,554.0 J
	Cu	0.0 J

Table 8.1. *Thermodynamic data of copper-based compounds*

8.3.2. *Step 2: choice of hydrated forms*

For each degree of oxidation, we need to choose the most stable degree of hydration. Indeed, the hydration reaction is neither an acid–base reaction nor a redox reaction, so there can be no distinctive domain for different hydrated forms.

In the case of copper, the problem lies in the choice between the two forms CuO and $Cu(OH)_2$.

To make this choice, we need only calculate the standard Gibbs energy associated with the hydration reaction. The value of –7,490 J per mole of CuO shows that the oxide is the stable form at the chosen temperature at a standard water vapor pressure of 1 bar. Therefore, we choose cupric oxide CuO to plot the diagram.

8.3.3. Step 3: study by degrees of oxidation of acid–base reactions; construction of the situation diagram

For each degree of oxidation, we shall determine the possible acid–base couples, and calculate the corresponding pK_a on the basis of the standard chemical potentials. For each couple, the domain of predominance (or of stability for a pure solid form) of the acidic form is for pH levels less than pK_a, whereas the basic form is predominant (stable) for higher pH values. Thus, we can determine the domains of pH of the predominance (or stability) of the corresponding acids and bases. In doing so, we obtain what is known as the *situation diagram*.

pH	0	5.16	6.95	7	12.83	13.5
Degree 2	Cu^{2+}			CuO	$HCuO_2$	CuO_2^{2-}
Degree 1	Cu^+			Cu_2O		
Degree 0				Cu		

Table 8.1. *Situation diagram for a concentration of 10^{-6}*

With copper at the degree of oxidation +1, there is only one possible reaction for the couple Cu^+/Cu_2O, which is written:

$$2Cu^+ + 3H_2O = Cu_2O + 2H_3O^+ \quad [8R.8]$$

At equilibrium, this reaction obeys the relation:

$$\log\left[Cu^+\right] = -0.84 - pH \quad [8.28]$$

pH	0	8.16	8.95	10.83	13.15
Degree 2	Cu^{2+}	CuO		$HCuO_2^-$	CuO_2^{2-}
Degree 1	Cu^+		Cu_2O		
Degree 0			Cu		

Table 8.2. *Situation diagram for a concentration of 10^{-8}*

At the degree of oxidation +2, in view of the amphoteric nature of CuO, we see the reaction with CuO playing the role of a base:

$$Cu^{2+} + 3H_2O = CuO + 2H_3O^+ \quad [8R.9]$$

That reaction, at equilibrium, must satisfy the relation:

$$\log\left[Cu^{2+}\right] = 7.89 - 2\,pH \quad [8.29]$$

and because of the acidic role of the same oxide, we have the couple:

$$CuO + 2H_2O = HCuO_2^- + H_3O^+ \quad [8R.10]$$

which obeys the condition:

$$\log\left[HCuO_2^-\right] = -18.83 + pH \quad [8.30]$$

or the equilibrium:

$$CuO + 3H_2O = CuO_2^{2-} + 2H_3O^+ \quad [8R.11]$$

which satisfies:

$$\log\left[CuO_2^{2-}\right] = -31.98 + 2\,pH \quad [8.31]$$

Finally, the equilibrium between CuO_2^{2-} and $HCuO_2^-$ gives us:

$$HCuO_2^- + H_2O = CuO_2^{2-} + H_3O^+ \quad [8R.12]$$

with the relation:

$$\log\frac{\left[CuO_2^{2-}\right]}{\left[HCuO_2^-\right]} = -13.15 + pH \quad [8.32]$$

An overview of these calculations is given in Tables 8.2 and 8.3, respectively for the plot concentrations $C = 10^{-6}$ and $C = 10^{-8}$ moles per liter of the element copper in solution. These tables show the pH domains in which each form is predominant or stable for the solid phases.

From these tables, we can immediately deduce the redox couples which need to be taken into account. For example at pH = 0, we take account, at the concentration of 10^{-6} moles/liter (Table 8.2), of the couples Cu/Cu^{+} and Cu^{+}/Cu^{2+}, and at pH = 7 it would be the couples CuO/Cu_2O and Cu/Cu_2O.

8.3.4. *Step 4: elimination of unstable species by dismutation*

Not all species included in tables such as Table 8.2 or 8.3 will necessarily have a domain of stability (or predominance) in the final diagram, because some of them may be unstable because of dismutation. It is helpful, at each pH, to calculate the redox voltages for each of the couples at the chosen concentration. In practice, it is often sufficient to perform the calculation for pH = 0 and one or two other values of the pH.

Thus, in the case of copper (see section 8.3.2), in the vicinity of pH = 0, we only need to consider the following couples:

for $C = 10^{-6}$ mol/liter:

$$Cu^{+} + e = Cu \qquad [8R.13]$$

where $e_0 = 0.52 - 0.06 \times 6 = 0.16\, V$

$$Cu^{2+} + e = Cu^{+} \qquad [8R.14]$$

where $e_0 = 0{,}15\, V$

For $C = 10^{-8}$ mol/liter:

$$Cu^{+} + e = Cu \qquad [8R.15]$$

where $e_0 = 0.52 - 0.06 \times 8 = 0.04\, V$

$$Cu^{2+} + e = Cu^{+} \qquad [8R.16]$$

where $e_0 = 0.15\, V$

The result of this is that at pH = 0, for a total concentration of 10^{-6} mol/liter, copper Cu^{+} dismutes and has no domain in the diagram, whereas for the total concentration of 10^{-8} mol/liter, that ion is stable.

At 10^{-6} mol/liter, we have the equilibrium:

$$Cu^{2+} + e = Cu \qquad [8R.17]$$

where $e_0 = 0.19\ V$

Similarly, we could verify that at pH = 8, Cu_2O does not dismute for the two concentrations examined here, with the couples:

$$Cu_2O + 2H_3O^{+} + 2e = 2Cu + 3H_2O \qquad [8R.18]$$

where $e^{0} = 0.47\ V$

$$2Cu^{2+} + 3H_2O + 2e = Cu_2O + 2H_3O^{+} \qquad [8R.19]$$

where $e^{0} = 0.203\ V$

For both concentrations, we need to take account of the equilibria:

$$2CuO + 2H_3O^{+} + 2e = Cu_2O + 3H_2O \qquad [8R.20]$$

where $e^{0} = 0.67\ V$

$$2HCuO_2^{-} + 4H_3O^{+} + 2e = Cu_2O + 7H_2O \qquad [8R.21]$$

where $e^{0} = 1.78\ V$

$$2CuO_2^{2-} + 6H_3O^{+} + 2e = Cu_2O + 9H_2O \qquad [8R.22]$$

where $e^{0} = 2.56\ V$

$$2CuO_2^{2-} + 4H_3O^{+} + 2e = 2CuO + 6H_2O \qquad [8R.23]$$

where $e^{0} = 1.51\ V$

Now with all this information at our fingertips, we can begin to plot the diagram.

8.3.5. *Step 5: plotting the e/pH diagram*

We plot the diagram little by little, starting at the value pH = 0.

In Figures 8.5 and 8.6, the numbers of the lines correspond to the equilibrium bearing the same number – for instance, line 17 corresponds to the equilibrium [8R.17].

At pH = 0, we plot the horizontal lines representing the redox equilibria without the involvement of hydrogen. Those horizontals are limited by acid–base reactions. Thus, we obtain (see Figures 8.5 and 8.6) the horizontal lines 14, 15 and 17 corresponding to the equilibria [8R.14], [8R.15] and [8R.17].

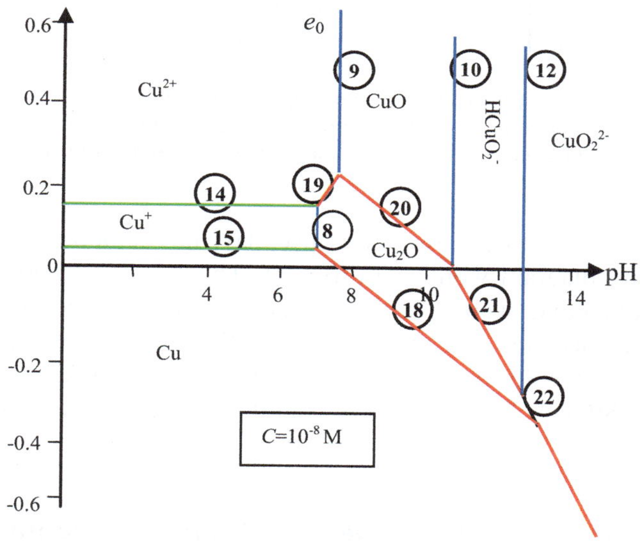

Figure 8.5. *Diagram for copper at the concentration of 10^{-8} moles/l. For a color version of the figure, see www.iste.co.uk/soustelle/equilibria.zip*

The first two, in Figure 8.5, are limited by the vertical line 8, representing the equilibrium [8R.8].

The intersections of the redox lines with the verticals define triple points through which the next redox equilibrium must pass. Thus, in Figure 8.6, we encounter the point T_1 first, followed by point T_2, then point T_3, point T_4 and finally point T_5.

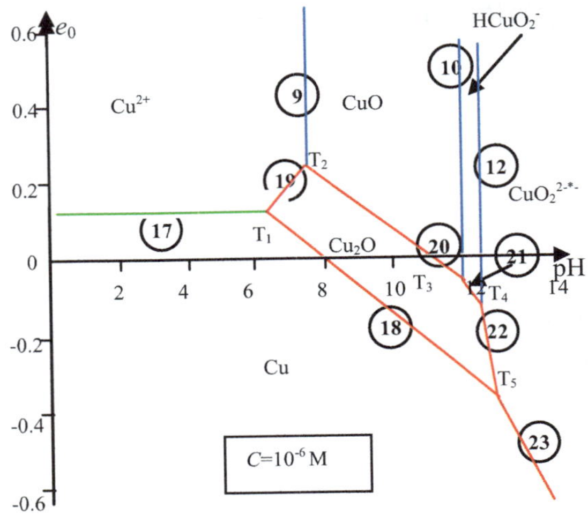

Figure 8.6. *Diagram for copper at the concentration of 10^{-6} moles/l. For a color version of the figure, see www.iste.co.uk/soustelle/equilibria.zip*

NOTE.– If we were to choose not to use, say, the compound $HCuO_2^-$, the corresponding diagram would be different. The zone corresponding to the excluded compound would no longer appear, a new equilibrium would be taken into account and the equilibria [8R.10] and [8R.12] would be eliminated. We would then see the appearance of a boundary linked to the equilibrium [8R.11]. The domains of stability of CuO and predominance of CuO_2^- would therefore be extended.

8.4. Diagram for water superposed on the diagram for an element

When we construct the diagram for copper, for instance, it is, in fact, the fait diagram for the copper/water system. The water still needs to be stable in the *e*/pH region being explored. In fact, water plays a part in two redox couples:

$$3H_2O = \frac{1}{2}O_2 + 2H_3O^+ + 2e \qquad [8R.24]$$

where $e = 1.23 - 0.06\,\text{pH}$

$$H_3O^+ + e = \frac{1}{2}H_2 + H_2O \qquad [8R.25]$$

where $e = -0.06\,\text{pH}$

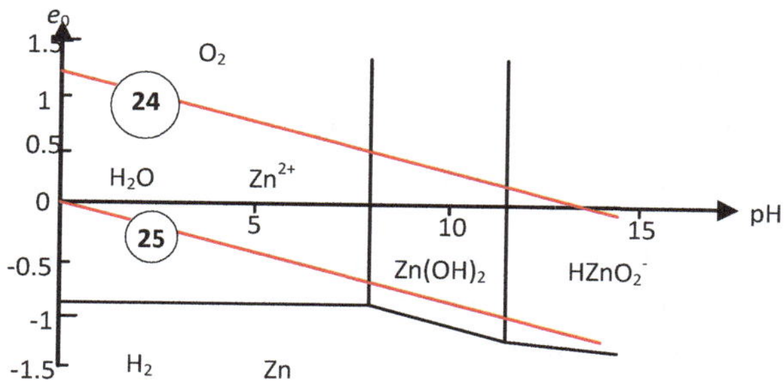

Figure 8.7. *Simplified diagram for zinc superposed on the diagram for water. For a color version of the figure, see www.iste.co.uk/soustelle/equilibria.zip*

Figure 8.6, for a concentration of 10^{-6} mol/liter, shows the simplified diagram for zinc, upon which are superposed the lines 24 and 25, delimiting the domain of stability of water.

Finally, the diagram for zinc (Figure 8.7) in an aqueous medium will contain only the two vertical lines, and the metal zinc will have no domain of stability.

NOTE.– In the diagram for water, account has not been taken of hydrogen peroxide H_2O_2 and ozone O_3. Calculations show that hydrogen peroxide is unstable and suffers dismutation, and therefore has no domain of stability. Ozone, meanwhile, can only be obtained in a mixture with oxygen by raising the electrode to a very high voltage, which is difficult to achieve with zero current.

8.5. Immunity, corrosion and passivation

Potential/pH diagrams of metal/water systems are very widely used for studying aqueous corrosion of metals. They are plotted at a very low total concentration (around 10^{-6} mol/liter), because corrosion is a slow phenomenon and gives rise only to low concentrations of species in solution.

From the standpoint of corrosion, potential/pH diagrams show three categories of domains of stability (or predominance).

– the domains of stability of the metal, wherein the metal is not attacked, known as the domain of immunity;

– the domains of corrosion, wherein the ions are predominantly present, and therefore the metal is attacked;

– domains of "passivation", in which we see stable solid phases which protect the metal from external corrosion.

NOTE.– In a domain of passivation, the diagram yields no information about the actual passivity of the metal, because we are then dealing with problems linked to both the texture of the solid and the kinetics of corrosion – both fields in which thermodynamics offers no information.

8.6. Potential/pX (e/pX) diagrams

As we have seen (in section 8.1.1), e/pH diagrams are plotted using specific conventions and in very precise conditions. If any one of those conditions is not respected, we need to start again and plot the diagram anew. If, for example, a metallic electrode is no longer a pure metal, but instead an alloy whose concentration (equal to its activity) [X], a third variable will come into play, and we shall then have three-dimensional e/pH/pX diagrams, where pX denotes -log [X].

It is sometimes interesting, and more easily operable, to plot such a diagram at a constant pH, in which case we obtain the two-dimensional e/pX diagram.

The same would be true if we were to add a compound which is likely to form complexes with the ions in the system. Such would be the case, for instance, with the diagram of silver in the presence of ammonium ions, which would again show a new variable – the concentration of ammonium ions – and two new compounds – the complexes $[Ag(NH_3)]^+$ and $[Ag(NH_3)_2]^+$.

By way of example, let us look at the simple case of the diagram for mercury in the presence of thiocyanate ions (SCN^-) and examine the corresponding e/pSCN diagram.

The introduction of the thiocyanate ions may trigger the precipitation of mercury thiocyanate $Hg_2(SCN)_2$ and the formation of $[Hg(SCN)_4]^{2-}$ ions.

Figure 8.8 shows the corresponding diagram with a plot concentration of 1 mole/l (i.e. at pH = 0).

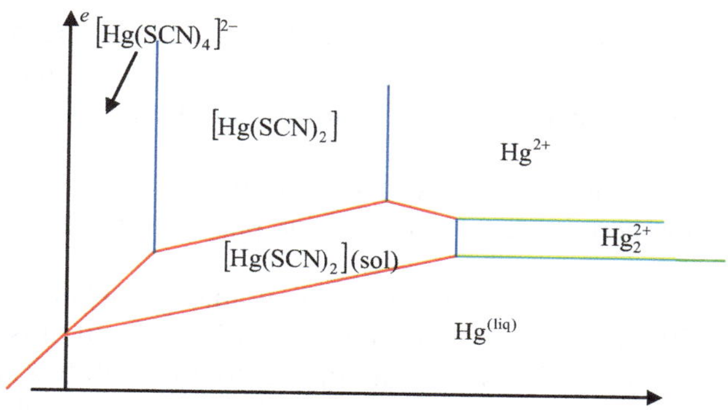

Figure 8.8. *The e/pSCN diagram for mercury for a concentration of 1mole i.e. (ph=0). For a color version of the figure, see www.iste.co.uk/soustelle/equilibria.zip*

Here, we see a solid zone which is no longer an oxide or a hydroxide, but mercury thiocyanate $Hg_2(SCN)_2$, and we note that for strong concentrations of thiocyanate, mercury(I) undergoes dismutation into liquid metal mercury and the complex $[Hg(SCN)_4]^{2-}$. There is no longer any domain of stability for the Hg_2^{2+} ion.

The use of such *e*/pX diagrams is entirely comparable to that of the *e*/pH diagrams we saw above.

8.7. Potential/acidity diagrams in a molten salt

For acceptations of acid–base couples other than those linked to the exchanges of protons, we can obviously plot and represent e/acidity diagrams that are absolutely similar to the e/pH diagrams for protic solvents. For example, on the basis of the oxo-acidity, we can plot e/pO^{2-} diagrams.

Of course, such diagrams are limited by oxidations and reductions linked to the solvent which limit its domain of useful potential.

The limits of the domain of pH of the solvent also appear, so in the molten solvent KCl-LiCl at 450°C, for example, the limit of pO^{2-} is obviously the precipitation of alkali oxides.

Thus, we define the window between the acidity zone and the potential zone, within which the diagram makes sense.

As is the case with e/pH diagrams, we choose a total concentration. The different domains thus delimited define zones of predominance for the species in solution and zones of stability for the solid species.

The same information can be gleaned from the diagram, and in particular, all the information pertaining to corrosion and passivation.

Figure 8.9 shows the example of the diagram for nickel in the solvent KCl-LiCl at 450°C, at a concentration of 10^{-6} mol/l. In this figure, we distinguish two zones of passivation, linked to the stability of the solid species NiO and NiO_2, and the zone of immunity of metal nickel.

It is noteworthy that a domain of stability of nickel at the degree of oxidation +4 (NiO_2) appears, though that form does not exist in aqueous solution.

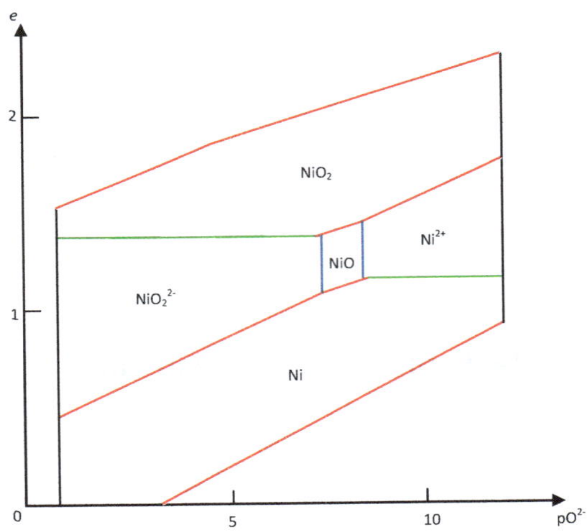

Figure 8.9. *The e/pO^{2-} diagram for nickel in the mixture KCl/LiCl at 450°C for a concentration of 10^{-6}mol/l. For a color version of the figure, see www.iste.co.uk/soustelle/equilibria.zip*

Appendix

Activities in Ionic Solutions

In ionic solutions, it is preferable to use convention (II) – the infinitely-dilute solution – or convention (III) – the 1 mol/l reference solution for each component. Thus, we need to know how to link the activity coefficients and the equilibrium constants for the two conventions with one another.

For a given solution, let us assign the values relative to the solvent with the index 0 and those relative to the solutes with the index s. Consider a reaction which is written:

$$0 = \sum_i \nu_i \mathrm{A}_i \qquad \text{[AR.1]}$$

The components i are those which take part in the reaction. They may be only solutes, or solutes and the solvent, or even include pure solid phases.

The equilibrium constant of this reaction is expressed, depending on the chosen convention, by the following two expressions:

– based on the molar fractions in reference (II):

$$K_r^{(\mathrm{II})} = \prod_s x_s^{\nu_s} \left[\gamma_s^{(\mathrm{II})}\right]^{\nu_s} = \exp - \frac{\Delta\mu_s^{\infty}}{\mathrm{R}T} \qquad \text{[A.1]}$$

The reference solution is the infinitely-dilute solution for all components.

– based on the concentrations in reference (III):

$$K_r^{(\mathrm{III})} = \prod_s C_s^{\nu_s} \left[\gamma_s^{(\mathrm{III})}\right]^{\nu_s} = \exp - \frac{\Delta\mu^{0(\mathrm{III})}}{\mathrm{R}T} \qquad \text{[A.2]}$$

The reference solution is that which contains 1 mol/l of each solute.

Only the terms relative to the solutes (and occasionally gases) are taken into account in the product of the second term in relations [A.1] and [A.2], with the exception of solvents in a dilute solution and pure solid phases. On the other hand, in the third term in those relations, the values $\Delta\mu^{\infty}$ and $\Delta\mu^{0(\mathrm{III})}$ contain the contributions of all the reagents, including the term relative to the solvent if it participates in the reaction and that of the solution, whether dilute or otherwise. Obviously, the same is true of the equilibrium constants $K_r^{(\mathrm{II})}$ and $K_r^{(\mathrm{III})}$.

A.1. Concentrations and molar fractions

By definition of molar fractions, we can write the ratio:

$$\frac{x_s}{x_0} = \frac{n_s}{n_0} \qquad [A.3]$$

However, the volume of the solution is given, as a function of the partial molar volumes of the components, by:

$$V = n_0\overline{V_0} + \sum_s n_s\overline{V_s} \qquad [A.4]$$

$\overline{V_0}$ and $\overline{V_s}$ respectively denote the partial molar volume of the solvent and that of the solute s.

Hence, by using the relation of definition of the concentration, we obtain:

$$C_s = \frac{n_s}{V} = \frac{n_s}{n_0\overline{V_0} + \sum_s n_s\overline{V_s}} = \frac{x_s}{x_0\overline{V_0} + \sum_s x_s\overline{V_s}} \qquad [A.5]$$

However, if the solution is sufficiently dilute and if v_0^0 denotes the molar volume of the pure solvent, we can write:

$$\sum_s x_s\overline{V_s} << x_0\overline{V_0} \cong \overline{V_0} \cong v_0^0 \qquad [A.6]$$

The molarity can therefore be written, approximately, as:

$$C_s \cong \frac{x_s}{v_0^0} \qquad [A.7]$$

The molar volume of solvent practically depends only on the temperature for a liquid or a solid.

A.2. Relation between the activity coefficients expressed in conventions (II) and (III)

Let us examine the relation between the activity coefficient of a component, expressed on the basis of the dilute solution reference (II), and that found using the molar solution convention (III).

We write that the chemical potential does not depend on the reference state, which is expressed by the equality:

$$\mu_s = \mu_s^{0(III)} + \mathrm{R}T \ln \gamma_s^{(III)} C_s \cong \mu_s^{\infty} + \mathrm{R}T \ln \gamma_s^{(II)} x_s \qquad [A.8]$$

Thus, we can identify and write:

$$\mu_s^{0(III)} = \mu_s^{\infty} \qquad [A.9]$$

and:

$$\gamma_s^{(III)} C_s = \gamma_s^{(II)} x_s \qquad [A.10]$$

Hence, the chemical potential at infinite dilution is equal to that at the concentration of 1 mole per liter, and the product of the activity coefficient in convention (II) by the molar fraction is equal to the product of the activity coefficient in convention (III) by the concentration expressed in moles per liter. In other words, the activities expressed in the two conventions are identical.

If the solution is very dilute, from relations [A.10] and [A.7], we deduce:

$$\gamma_s^{(II)} = \frac{\gamma_s^{(III)}}{v_0^0} \qquad [A.11]$$

The molar volume of the solvent v_0^0 is expressed in liters per mole.

A.3. Relation between the equilibrium constants expressed in conventions (II) and (III)

Because the chemical potentials of the reference states (II) and (III) are equal (see relation [A.9]), the equilibrium constant is the same in both cases:

$$K_r^{(III)} = K_r^{(\mathrm{II})} \tag{A.12}$$

If we now combine relations [A.1] [A.2] and [A.12], we obtain the following relation (only the solutes come into play here):

$$\prod_s \left(\gamma_s^{(\mathrm{III})} C_s\right)^{\nu_s} = \prod_s \left(\gamma_s^{(\mathrm{II})} x_s\right)^{\nu_s} = K_r^{(\mathrm{II})} \tag{A.13}$$

If the reference concentration is 1 mol/l and we consider only sufficiently-dilute solutions – i.e. solutions such that $\gamma_s^{(\mathrm{II})} = 1$ – then in light of relation [A.7], we find:

$$\prod_s (C_s)^{\nu_s} = K_r^{(\mathrm{II})} \left(v_0^0\right)^{\sum_s \nu_s} = \left(v_0^0\right)^{\sum_s \nu_s} \prod_s (x_s)^{\nu_s} \tag{A.14a}$$

NOTE.– If the reaction occurs with a constant number of moles ($\sum \nu_s = 0$), we can treat the activities as being the same as the molar fractions and the concentrations, and write:

$$\prod_s (C_s)^{\nu_s} = \prod_s (x_s)^{\nu_s} = K_r^{(\mathrm{II})} \tag{A.14b}$$

Note that the term $\sum_s \nu_s$ in relation [A.14a] pertains only to species in solution, to the exclusion of solids and the solvent.

For ionic reactions in solution, studied in this volume, even if this condition does not appear to be respected, this stems from what is often an abbreviated formulation of the reaction, in which the solvent is not always explicitly written. Take the example of the dissociation of ethanoic acid. Often, this reaction is written in the form:

$$CH_3COOH = CH_3COO^- + H^+ \tag{AR.2}$$

The term $\sum_s \nu_s = 1$ is non-null. In actual fact, the complete equilibrium, which only occurs in the presence of water, should be written:

$$CH_3COOH + H_2O = CH_3COO^- + H_3O^+ \qquad [AR.3]$$

Thus, we can see that, in reality, $\sum_s \nu_s = 0$ and therefore, in spite of appearances, the reaction [AR.2] does indeed yield expression [A.14b].

The same is true for any acid–base couple or any redox couple for which the sum $\sum_s \nu_s$ is non-null: that sum takes the value of 0 and disappears when writing the overall reaction of that couple with a different couple.

Thus, in ionic reactions in a dilute solution, we can apply relation [A.13], or if the concentrations are sufficiently low, relation [A.14b].

A.4. Correspondence between conventions (I) and (II)

Remember that the activity coefficient of a solute in reference (I) is that obtained by choosing the pure substance as a reference state. That activity coefficient in reference (I) and that in reference (II) (the infinitely-dilute solution) are interlinked, because the chemical potential of the solute does not depend on the convention chosen. Thus, we can write:

$$\mu_s = g_s^0 + RT \ln \gamma_s^{(I)} x_s = \mu_s^\infty + RT \ln \gamma_s^{(II)} x_s \qquad [A.15]$$

If K_{iH} denotes Henry's constant, defined by the relation:

$$\ln K_{iH} = \frac{\mu_s^\infty - g_s^0}{RT} \qquad [A.16]$$

By comparing relations [A.15] and [A.16], we can deduce:

$$\ln \frac{\gamma_s^{(I)}}{\gamma_s^{(II)}} = \frac{\mu_s^\infty - g_s^0}{RT} = \ln K_{iH} \qquad [A.17]$$

Thus, the constant K_{iH} links the activity coefficients expressed, for a given solute, in both conventions: the pure-substance reference and the dilute-solution reference. This constant does not depend on the composition of the solution. However, by way of the chemical potentials of the reference states, it does depend on the values of the intensive variables (pressure, temperature, and as seen in Chapter 2, the solvent, etc.).

The equilibrium constant in reference (I) is written:

$$K_r^{(\mathrm{I})} = \exp-\frac{\Delta g^0}{\mathrm{R}T} = \prod_s \left(\gamma_s^{(\mathrm{I})} x_s\right)^{\sum_s \nu_s} \qquad \text{[A.18]}$$

Using relation [A.17], we obtain:

$$K_r^{(\mathrm{II})} = \frac{K_r^{(\mathrm{I})}}{\prod_s K_{sH}^{\nu_s}} \qquad \text{[A.19]}$$

NOTE.– For a perfect solution, where the Henry constants for the different components are all equal to 1, the equilibrium constants $K_r^{(II)}$ and K_r are treated as one and the same thing. As, according to relation [A.12], the equilibrium constants in reference (II) and reference (III) are identical, it follows that for a perfect solution, all the equilibrium constants are identical.

A.5. Debye and Hückel's model for ionic solutions

We know that it is impossible to measure the activity coefficient of an ion, because it is not possible to isolate an ion in a solution because of the need to preserve electrical neutrality. On the other hand, it is possible to determine the activity coefficient in reference (I) using models. Models of ionic solutions (which are described in detail in Volume 3 of this set [SOU 15c]), are all based on Debye and Hückel's model, which takes account of Coulomb force attractions between the ions in the calculation of the excess Gibbs energy. More sophisticated models complete the description of the solution, by also taking account of the interactions over short distances between the molecules, between the molecules and ions, and between the ions themselves, based on London forces, for instance.

With the ionic equilibria, working on dilute solutions, we content ourselves with the simplified form of the Debye–Hückel equation for the activity coefficient of an ion. This form is written:

$$\ln \gamma_s = -Bz_s^2\sqrt{I} \qquad \text{[A.20]}$$

I is the ionic strength, whose definition is:

$$I = \sum_i C_i z_i^2 \qquad [A.21]$$

This sum is extended to all ions present in the solution.

In fact, the constant B in relation [A.20] depends on the solvent.

For water, at the temperature of 25°C, we have:

$$\begin{cases} B = 0.511 \\ D = 78.5 \\ B' = 355.4 \end{cases}$$

Bibliography

[BER 97] BERNACHE-ASSOLLANT D., COURNIL M., *Equilibres en solution*, Masson, Paris 1997.

[BES 84] BESSON J., *Précis de thermodynamique et de cinétique électrochimiques*, Ellipses, Paris, 1984.

[BJE 29] BJERRUM N., "Neuere Anschauungen über Elektrolyte" *Berichte d.D. Gesellshaft*, vol. 62, pp. 1091–1103, 1929.

[DEM 73] DEMANGE H., GERMAIN G., NOTIN M., *Comprendre et appliquer la thermodynamique*, Masson & Cie, Paris, 1973.

[DEN 68] DENBIGH K., *The principles of chemical equilibrium*, University Press, Cambridge, 1968.

[DÉP 94] DÉPORTES C., DUCLOT M., FABRY P. *et al.*, *Electrochimie des solides*, EDP Sciences, 1994.

[EMS 51] EMSCHWILLER G., *Chimie Physique*, Presses Universitaire de France, Paris, 1951.

[FOW 49] FOWLER R., GUGGENHEIM E.A., *Statistical Thermodynamics*, Cambridge University Press, London, 1949.

[FUO 58] FUOSS R.M., "Conductance of Ionopheres", *J. Amer. Chem. Soc.*, vol. 80, p. 3163, 1958.

[GEL 97] GELLINGS P.J., BOUWMEESTER H.J.M., *Solid State Electrochemistry*, CRC Press Inc, 1997.

[GOK 96] GOKCEN N.A., REDDY R.G., *Thermodynamics*, Plenum Press, New York, 1996.

[GUE 09] GUETTA V., FOULETIER J., TAXIL P., *Sels fondus à haute température*, Presses Universitaires Polytechniques Romandes, Lausanne, 2009.

[HAM 32] HAMMETT L.P., DEYRUP J., "A series of simple basic indicators. II. Some applications to solutions in formic acid", *J. Amer. Chem. Soc.*, vol. 54, p. 272, 1932.

[IMP 86] IMPEY R.W., "Computer simulation of water and ionic solutions: what can we learn?", in CICOTTI G., HOOVER W.G (eds), *Molecular Dynamics Simulation of Statistical Mechanical Systems*, North-Holland, Amsterdam 1986.

[INF 06] INFELTA P., GRAETZEL M., *Thermodynamique, Principes et Applications*, BrownWalker Press 2006.

[JOB 28] JOB P., "Formation and stability of inorganic complexes", *Analytical Chemistry*, vol. 9, no. 10, pp. 113–203, 1928.

[KLO 08] KLOTZ I.M., ROSENBERG R.W., *Chemical Thermodynamics: Basic Concepts and Methods*, 7th ed., John Wiley & Sons, 2008.

[KRÖ 74] KRÖGER F.A., *The Chemistry of Imperfect Crystals*, North-Holland Publishing Company, Amsterdam, 1974.

[LAL 06] LALAUZE R., *Physico-chimie des interfaces solide-gaz 1*, Hermès–Lavoisier, Paris, 2006.

[LE 05] LE GORREC B., Thermodynamique et cinétique électrochimique expérimentale, available at: www.minatec.grenoble-inp.fr/Cours-A.pdf, 2005.

[MCQ 99] MCQUARRIEN D.A., SIMON J.D., *Molecular Thermodynamics*, University Science Book, Sausalito, US, 1999.

[PAC 63] PACAULT A., *Eléments de thermodynamique statistique*, Masson, Paris, 1963.

[PRI 46] PRIGOGINE I., DEFAY R., *Thermodynamique chimique*, Dunod, Paris, 1946.

[ROS 85] ROSENBERG R.M., HOBBS E.V., "The spreadsheet", *Chem. Educ.*, vol. 62, pp. 140–141, 1985.

[SOR 81] SORENSEN O.T., *Non-stoichiometric Oxides*, Academic Press, New York, 1981.

[SOU 68] SOUCHAY P., *Thermodynamique chimique*, Masson, Paris, 1968.

[SOU 90] SOUSTELLE M., *Modélisation macroscopique des transformations physico-chimiques*, Masson, Paris, 1990.

[SOU 93] SOUSTELLE M., "Défauts ponctuels dans les solides", in BERNACHE-ASSOLLANT D. (ed.), *Chimie–physique du frittage*, Hermès, Paris, 1993.

[SOU 10] SOUSTELLE M., *Handbook of Heterogeneous Kinetics*, ISTE Ltd., London, and John Wiley & Sons, New York, 2010.

[SOU 11] SOUSTELLE M., *An Introduction to Chemical Kinetics*, ISTE, London and John Wiley & Sons, New York, 2011.

[SOU 15a] SOUSTELLE M., *Phase Modeling Tools – Application to Gases*, (Volume 1 of the *Chemical Thermodynamics Set*), ISTE, London and John Wiley & Sons, New York, 2015.

[SOU 15b] SOUSTELLE M., *Modeling of Liquid Phases* (Volume 2 of the *Chemical Thermodynamics Set*), ISTE, London and John Wiley & Sons, New York, 2015.

[SOU 15c] SOUSTELLE M., *Thermodynamic Modeling of Solid Phases*, (Volume 3 of the *Chemical Thermodynamics Set*), ISTE, London and John Wiley & Sons, New York, 2015.

[SOU 15d] SOUSTELLE M., *Chemical Equilibria* (Volume 4 of the *Chemical Thermodynamics Set*), ISTE, London and John Wiley & Sons, New York, 2015.

[SOU 16] SOUSTELLE M., *Phase Transformations* (Volume 5 of the *Chemical Thermodynamics Set*), ISTE, London and John Wiley & Sons, New York, 2016.

[TRÉ 71] TRÉMILLON B., *La chimie en solvants non aqueux*, PUF, Paris 1971.

Index

A, B, C

D, E

F, G

H, I, J

K, L, M

N, O, P

Q, R, S

T, V

Printed and bound by CPI Group (UK) Ltd, Croydon, CR0 4YY
16/09/2026
14963059-0001